A Textbook on

Linear Differential Equation

Engineering Mathematics

(Theory & Solved Examples)

By

M. D. PETALE

Purpose of this Book

The purpose of this book is to supply lots of examples with details solution that helps the students to understand each example step wise easily and get rid of the College assignments phobia.

It is sincerely hoped that this book will help and better equipped the higher secondary students to prepare and face the examinations with better confidence. I have endeavored to present the book in a lucid manner which will be easier to understand by all the engineering students.

About the Book

According to many streams in engineering course there are different chapters in Engineering Mathematics of the same year according to the streams. Hence students faced problem about to buy Engineering Mathematics special book that covered all chapters in a single book.

That's reason student needs to buy many books to cover all chapters according to the prescribed syllabus. Hence need to spend more money for a single subject to cover complete syllabus.

So here good news for you, your problem solved.

I made here special books according to chapter wise, which helps to buy books according to chapters and no need to pay extra money for unneeded chapters that not mentioned in your syllabus.

About the Author

M. D. PETALE

PREFACE

Dear Students,

It gives me great pleasure to present to you this book on A Textbook on **"Linear Differential Equation"** of Engineering Mathematics presented specially for you.

Many books have been written on Engineering Mathematics by different authors and teachers, but majority of the students find it difficult to fully understand the examples in these books.

Also, the Teachers have faced many problems due to paucity of time and classroom workload. Sometimes the college teacher is not able to help their own student in solving many difficult questions in the class even though they wish to do so.

Keeping in mind the need of the students, the author was inspired to write a suitable text book providing solutions to various examples of **"Linear Differential Equation"** of Engineering Mathematics.

It is hoped that this book will meet more than an adequately the needs of the students they are meant for. I have tried our level best to make this book error free.

Any suggestions for the improvement of the book would be most welcome and gratefully acknowledged.

Author
M. D. PETALE

A Textbook on

Linear Differential Equation

Author

M.D. PETALE

Website: http://www.mytechnofuture.com

Email: mdpetale1@gmail.com

Type Editing and Setting

M.D. PETALE

Price: Rs. 250/-

CONTENTS

Linear Differential Equation

1 Introduction

The ordinary/linear differential equation with constant coefficient find their most important application in the study of electrical, mechanical and other linear system. In fact such equations play a dominant role in unifying the theory of electrical and mechanical oscillatory system. Such equations are most important in the study of electromechanical vibration and equivalent electrical circuits and their engineering problems.

1.i Definition

An ordinary differential equation is a relationship between a real variable, (let us say x), a dependent variable (let us call this y), and (possibly many)derivatives of the dependent variable y with respect to x.

Thus the **general linear differential equation** of the nth order is of the form

$$p_0 \frac{d^n y}{dx^n} + p_1 \frac{d^{n-1} y}{dx^{n-1}} + p_2 \frac{d^{n-2} y}{dx^{n-2}} + \cdots + p_n y = X \qquad \ldots \ldots (1)$$

Where $p_0, p_1, p_2, \ldots, p_n$ and X are functions of x only or constant terms.

Linear differential equations with constant coefficients are of the form

$$\frac{d^n y}{dx^n} + k_1 \frac{d^{n-1} y}{dx^{n-1}} + k_2 \frac{d^{n-2} y}{dx^{n-2}} + \cdots + k_n y = X \qquad \ldots \ldots (2)$$

Where $k_1, k_2, \ldots, kn$ are constants only and X is function of x or constant.

2 Operator D

Denoting $\frac{d}{dx} = D$, $\frac{d^2}{dx^2} = D^2$, $\frac{d^3}{dx^3} = D^3$ etc.

So that $\frac{dy}{dx} = Dy$, $\frac{d^2 y}{dx^2} = D^2 y$, $\frac{d^3 y}{dx^3} = D^3 y$ etc.

The equation (2) above can be written in the symbolic form as

$$\left(D^n + k_1 D^{n-1} + k_2 D^{n-2} + \cdots + k_n\right) y = X$$

i.e. $f(D)y = x$

Where $f(D) = D^n + k_1 D^{n-1} + k_2 D^{n-2} + \cdots + k_n$

i.e. A polynomial in D.

Thus the symbol D stands for the operation of differentiation and can be much the same as an algebraic quantity i.e. $f(D)$ can be factorized by ordinary rules of algebra and the factor may be taken in any order.

Ex. $\dfrac{d^2y}{dx^2} + 2\dfrac{dy}{dx} - 3y = (D^2 + 2D - 3)y$

$\qquad\qquad\qquad = (D + 3)(D - 1)y \quad$ Or $\quad (D - 1)(D + 3)y$

3 Rules for finding the Complementary Function (C. F.)

To solve the equation $\dfrac{d^n y}{dx^n} + k_1 \dfrac{d^{n-1}y}{dx^{n-1}} + \cdots + k_n y = 0$

Where k's are constants.

The above equation in symbolic form is

$(D^n + k_1 D^{n-1} + \cdots + k_n)y = 0$

The symbolic equation equated to zero i. e.

$(\mathbf{D^n + k_1 D^{n-1} + \cdots + k_n}) = \mathbf{0}$ **is called the Auxiliary Equation (A. E.)**

i. e. $f(D) = 0$

Let, $\quad \alpha_1, \alpha_2, \alpha_3, \ldots \ldots \alpha_n$ be its roots.

Case I: All roots Real &Distinct (Different or Non repeated or Unequal)

If $D = \alpha_1, \alpha_2, \alpha_3, \alpha_4, \ldots$

Then, $\,$ C. F. $= C_1 e^{\alpha_1 x} + C_2 e^{\alpha_2 x} + C_3 e^{\alpha_3 x} + C_4 e^{\alpha_4 x} + \cdots$

Case II: All roots Real &Same (Repeated or Equal)

If $D = \alpha_1, \alpha_2, \alpha_3, \alpha_4, \ldots$ but $\alpha_1 = \alpha_2 = \alpha_3 = \alpha_4 = \alpha$ (say)

Then, $\,$ C. F. $= (C_1 + C_2 x + C_3 x^2 + \cdots)e^{\alpha x}$

Case III: All roots Complex &Distinct (Different or Non repeated or Unequal)

If $D = \alpha_1 + i\beta_1, \quad \alpha_2 + i\beta_2, \quad \ldots$

Then, $\,$ C. F. $= e^{\alpha_1 x}(C_1 \cos\beta_1 x + C_2 \sin\beta_1 x) + e^{\alpha_2 x}(C_3 \cos\beta_2 x + C_4 \sin\beta_2 x) + \cdots$

Case IV: All roots Complex &Same (Repeated or Equal)

If $D = \alpha_1 \pm i\beta_1, \alpha_2 \pm i\beta_2, \quad \alpha_3 \pm i\beta_3, \quad \ldots$

But $\alpha_1 \pm i\beta_1 = \alpha_2 \pm i\beta_2 = \alpha_3 \pm i\beta_3 = \alpha \pm i\beta$(say)

Then, $\,$ C. F. $= e^{\alpha x}[(C_1 + C_2 x + C_3 x^2 + \cdots)\cos\beta x + (C_4 + C_5 x + C_6 x^2 + \cdots)\sin\beta x]$

Note: $\quad$ C. F. contains arbitrary constants

$\qquad$ Where, $C_1, C_2, C_3 \ldots, C_4, C_5, C_6 \ldots$ are arbitrary constants.

4   Inverse operator

Definition: $\dfrac{1}{f(D)}$ X is that function of x, not containing arbitrary constants

which when operated upon by f(D) given X

$$\text{i.e. } f(D)\left\{\frac{1}{f(D)}X\right\} = X$$

Thus $\dfrac{1}{f(D)}$ X satisfies the equation $f(D)y = X$ and is, Therefore, its particular integral

Thus,   **P.I.** $= \dfrac{1}{f(D)}$ X

5   Rules for finding the Particular Integral (P. I.)

Type I: General method

$$X = e^{e^x}, \sin e^x, \cos e^x, \sec x, \operatorname{cosec} x, \tan x, \cot x, \log x, \frac{1}{x}, \frac{1}{1+x^2}$$

   i) $f(D) = D - a$

$$\therefore \frac{1}{D-a} X = e^{ax} \int e^{-ax} X \, dx$$

   ii) $f(D) = D + a$

$$\therefore \frac{1}{D+a} X = e^{-ax} \int e^{ax} X \, dx$$

Short cut methods: Type II to Type VI

Type II: $X = e^{ax}$

   i) $\dfrac{1}{f(D)} e^{ax} = \dfrac{e^{ax}}{f(a)}$,   provided $f(a) \neq 0$

   ii) $\dfrac{1}{(D-a)^r} e^{ax} = \dfrac{x^r}{r!} e^{ax}$

Type III: $X = \sin(ax + b)$ or $\cos(ax + b)$

   i) $\dfrac{1}{f(D^2)} \sin(ax + b) = \dfrac{\sin(ax + b)}{f(-1.a^2)}$,   provided $f(-1.a^2) \neq 0$

   ii) $\dfrac{1}{f(D^2)} \cos(ax + b) = \dfrac{\cos(ax + b)}{f(-1.a^2)}$,   provided $f(-1.a^2) \neq 0$

   iii) $\dfrac{1}{(D^2+a^2)^r} \sin(ax + b) = \left(\dfrac{-x}{2a}\right)^r \dfrac{1}{r!} \sin\left(ax + b + \dfrac{r\pi}{2}\right)$

   iv) $\dfrac{1}{(D^2+a^2)^r} \cos(ax + b) = \left(\dfrac{-x}{2a}\right)^r \dfrac{1}{r!} \cos\left(ax + b + \dfrac{r\pi}{2}\right)$

Type IV: $X = x^m$, m is a positive integer

$$\frac{1}{f(D)} x^m = \frac{1}{1 \pm z} x^m, \quad \text{where z is function of D} = (1 \pm z)^{-1} x^m$$

Note: 1) $(1 + z)^{-1} = 1 - z + z^2 - z^3 + \cdots$

2) $(1 - z)^{-1} = 1 + z + z^2 + z^3 + \cdots$

3) $(1 + z)^n = 1 + nz + \dfrac{n(n-1)}{2!} z^2 + \dfrac{n(n-1)(n-2)}{3!} z^3 + \cdots$

4) $(1 + z)^{-2} = 1 - 2z + 3z^2 - 4z^3 + \cdots$

5) $(1 - z)^{-2} = 1 + 2z + 3z^2 + 4z^3 + \cdots$

Type V: $X = e^{ax}.V$, **where V is a function of x**

$$\frac{1}{f(D)} e^{ax}.V = e^{ax} \frac{1}{f(D + a)} V$$

Type VI: $X = x.V$, where V is a function of x

$$\frac{1}{f(D)} x.V = x \frac{1}{f(D)} V - \frac{f'(D)}{[f(D)]^2} V$$

Special cases:

1) $X = a^x$

$$\frac{1}{f(D)} a^x = \frac{1}{f(D)} e^{\log a^x} = \frac{1}{f(D)} e^{x \log a} = \frac{e^{(\log a)x}}{f(\log a)} = \frac{a^x}{f(\log a)}$$

2) $X = x^m . \cos ax$ or $x^m \sin ax$

i) $\dfrac{1}{f(D)} x^m . \cos ax = \text{R.P. of } \dfrac{1}{f(D)} e^{iax} . x^m = \text{R.P. of } e^{iax} \dfrac{1}{f(D + ia)} x^m$

ii) $\dfrac{1}{f(D)} x^m . \sin ax = \text{I.P. of } \dfrac{1}{f(D)} e^{iax} . x^m = \text{I.P. of } e^{iax} \dfrac{1}{f(D + ia)} x^m$

3) $X = k$, where k is constant

$$\frac{1}{f(D)} k = k \frac{1}{f(D)} . 1 = k \frac{1}{f(D)} e^{0x} = k \frac{e^{0x}}{f(0)} = \frac{k}{f(0)}, \quad \text{provided } f(0) \neq 0$$

4) $\dfrac{1}{D} X = \displaystyle\int X \, dx$

Note: When X is any other function of x.

P.I. $= \dfrac{1}{f(D)} X$; where $f(D) = (D - m_1)(D - m_2)$

Resolving into partial fraction

$$\frac{1}{f(D)} = \frac{A_1}{D - m_1} + \frac{A_2}{D - m_2} + \cdots + \frac{A_n}{D - m_n}$$

$$\text{P.I.} = \left[\frac{A_1}{D - m_1} + \frac{A_2}{D - m_2} + \cdots + \frac{A_n}{D - m_n}\right] X$$

$$= A_1 \frac{1}{D - m_1} X + A_2 \frac{1}{D - m_2} X + \cdots + A_n \frac{1}{D - m_n} X$$

$$\text{P.I.} = A_1 . e^{m_1 x} \int e^{-m_1 x} X \, dx + A_2 e^{m_2 x} \int e^{-m_2 x} X \, dx + \cdots + A_n e^{m_n x} \int e^{m_n x} X \, dx$$

5.i Examples on C.F.

Case I: All roots are Real &Distinct

Example 1: Solve $\dfrac{d^2 x}{dt^2} + 5\dfrac{dx}{dt} + 6x = 0$, given $x(0) = 0$; $\dfrac{dx(0)}{dt} = 15$

Solution: Given equation, $\dfrac{d^2 x}{dt^2} + 5\dfrac{dx}{dt} + 6x = 0$

In symbolic form $(D^2 + 5D + 6)x = 0$

Its A. E. is $f(D) = 0$, $D^2 + 5D + 6 = 0$

    i. e. $(D + 2)(D + 3) = 0$

$\therefore D = -2, -3$

$\therefore$ C. F. $= C_1 e^{-2t} + C_2 e^{-3t}$

and P. I. $= 0$

$\therefore$ C. S. $=$ C. F. $+$ P. I.

C. S. $= x = C_1 e^{-2t} + C_2 e^{-3t}$ ... ... (1)

    Diffirentiate w. r. t. 't'

$\therefore \dfrac{dx}{dt} = -2C_1 e^{-2t} - 3C_2 e^{-3t}$ ... ... (2)

When $t = 0$, $x = 0$ $\{\because$ given $x(0) = 0$

Equation (1) becomes, $0 = C_1 + C_2$ ... ... (3)

When $t = 0$, $\dfrac{dx}{dt} = 15$ $\left\{\because \dfrac{dx}{dt}(0) = 15\right.$

Equation(2) becomes, $15 = -2C_1 - 3C_2$ ... ... (4)

Equation (3) $\times 2$ $2C_1 + 2C_2 = 0$

Equation (4) $\underline{-2C_1 - 3C_2 = 15}$

    Adding $-C_2 = 15$ $\therefore C_2 = -15$

Put $C_2 = -15$ in equation (3) we get, $0 = C_1 - 15$ $\therefore C_1 = 15$

$\therefore$ Equation (1) becomes, $x = 15 e^{-2t} - 15 e^{-3t}$

$\therefore x = 15 \left(e^{-2t} - e^{-3t}\right)$ ... Required solution

Case II: All roots are Real &Same

Example 2: Solve $\dfrac{d^2x}{dt^2} + 6\dfrac{dx}{dt} + 9x = 0$

Solution: Given equation, $\dfrac{d^2x}{dt^2} + 6\dfrac{dx}{dt} + 9x = 0$

In symbolic form $(D^2 + 6D + 9)x = 0$

Its A. E. is $f(D) = 0$, $D^2 + 6D + 9 = 0$

i. e. $(D + 3)^2 = 0$

$\therefore D = -3, -3$

$\therefore$ C. F. $= (C_1 + C_2 t)e^{-3t}$

and P. I. $= 0$

$\therefore$ C. S. $=$ C. F. $+$ P. I.

$\therefore$ C. S is $x = (C_1 + C_2 t)e^{-3t}$

Example 3: Solve $(D^4 - 4D^2 + 4)y = 0$

Solution: Given equation, $(D^4 - 4D^2 + 4)y = 0$

Its A. E. is $f(D) = 0$, $D^4 - 4D^2 + 4 = 0$

i. e. $(D^2 - 2)^2 = 0$; $(D^2 - 2)(D^2 - 2) = 0$; $(D^2 - 2) = 0$ or $(D^2 - 2) = 0$

$\therefore D^2 = 2, 2$

$\therefore D = \pm\sqrt{2}, \pm\sqrt{2}$

$\quad = \sqrt{2}, \sqrt{2}, -\sqrt{2}, -\sqrt{2}$

$\therefore$ C. F. $= (C_1 + C_2 x)e^{\sqrt{2}\,x} + (C_3 + C_4 x)e^{-\sqrt{2}\,x}$

and P. I. $= 0$

$\therefore$ C. S. $=$ C. F. $+$ P. I.

$\therefore$ C. S. is $y = (C_1 + C_2 x)e^{\sqrt{2}\,x} + (C_3 + C_4 x)e^{-\sqrt{2}\,x}$

Case III: All roots are Complex &Distinct

Example 4: Solve $\dfrac{d^4x}{dt^4} + 4x = 0$

Solution: Given equation in symbolic form $(D^4 + 4)x = 0$

Its A. E. is $f(D) = 0$, $D^4 + 4 = 0$

i. e. $(D^4 + 4D^2 + 4) - 4D^2 = 0$ ... Note down this step(Adjustment)

$\quad (D^2 + 2)^2 - (2D)^2 = 0$; $(D^2 + 2 + 2D)(D^2 + 2 - 2D) = 0$

$\therefore (D^2 + 2D + 2)(D^2 - 2D + 2) = 0$ $\{\because a^2 - b^2 = (a+b)(a-b)\}$

$\therefore D^2 + 2D + 2 = 0$ or $D^2 - 2D + 2 = 0$

Use quadratic formula $ax^2 + bx + c = 0$ $\therefore \ x = \dfrac{-b \pm \sqrt{b^2 - 4ac}}{2a}$

$\therefore D = \dfrac{-2 \pm \sqrt{(2)^2 - 4(1)(2)}}{2(1)}$ orD $= \dfrac{2 \pm \sqrt{(-2)^2 - 4(1)(2)}}{2(1)}$

$\therefore D = \dfrac{-2 \pm \sqrt{-4}}{2}$ or $\dfrac{2 \pm \sqrt{-4}}{2}$ $= \dfrac{-2 \pm 2i}{2}$ or $\dfrac{2 \pm 2i}{2}$

$\therefore D = -1 \pm i \ , \ 1 \pm i$

$\therefore$ C.F. $= e^{-t}(C_1 \cos t + C_2 \sin t) + e^t(C_3 \cos t + C_4 \sin t)$

and P.I. $= 0$

$\therefore$ C.S. $=$ C.F. $+$ P.I.

$\therefore$ C.S. isx $= e^{-t}(C_1 \cos t + C_2 \sin t) + e^t(C_3 \cos t + C_4 \sin t)$

Case IV: All roots are Complex &Same

Example 5: Solve $(D^2+1)^3 y = 0$,   where $D = \dfrac{d}{dx}$

Solution: Given equation, $(D^2+1)^3 y = 0$

Its A. E. is f(D) = 0, $(D^2+1)^3 = 0$; (D^2+1) (D^2+1) (D^2+1) $= 0$

i. e. $D^2 = -1, \ -1, \ -1$

   $\therefore D = \pm i, \ \pm i, \ \pm i$

$\therefore$ C. F. $= e^{0x}[(C_1 + C_2 x + C_3 x^2) \cos x + (C_4 + C_5 x + C_6 x^2) \sin x]$

and P. I. $= 0$

$\therefore$ C. S. $=$ C. F. $+$ P. I.

$\therefore$ C. S. is $y = (C_1 + C_2 x + C_3 x^2) \cos x + (C_4 + C_5 x + C_6 x^2) \sin x$

Mix Cases:

Example 6: Solve $(D^3 + D^2 + 4D + 4)y = 0$

Solution: Given equation, $(D^3 + D^2 + 4D + 4)y = 0$

Its A. E. is f(D) = 0, $D^3 + D^2 + 4D + 4 = 0$

$\therefore$ By synthetic division

$$
\begin{array}{r|rrrr}
-1 & 1 & 1 & 4 & 4 \\
   &   & -1 & 0 & -4 \\
\hline
   & 1 & 0 & 4 & 0
\end{array}
$$

$\therefore (D + 1)(D^2 + 4) = 0$; $(D + 1) = 0$ or $(D^2 + 4) = 0$; $D = -1$ or $D^2 = -4$

$\therefore D = -1$, $D = \pm 2i = 0 \pm 2i$

$\therefore$ C. F. $= C_1 e^{-x} + e^{0x}(C_2 \cos 2x + C_3 \sin 2x) = C_1 e^{-x} + (C_2 \cos 2x + C_3 \sin 2x)$

and P.I. $= 0$

$\therefore$ C.S. $=$ C.F. $+$ P.I.

$\therefore$ C.S. is $y = C_1 e^{-x} + C_2 \cos 2x + C_3 \sin 2x$ $\qquad \{\because e^0 = 1$

5.ii   Examples on P.I.

Type I: $X = e^{e^x}$, $\sin e^x$, $\cos e^x$, $\sec x$, $\operatorname{cosec} x$, $\tan x$, $\cot x$, $\log x$, $\dfrac{1}{x}$, $\dfrac{1}{1+x^2}$

Example 7: Find the P.I. of $(D^2 + 3D + 2)y = \sin e^x$

Solution: P.I. $= \dfrac{1}{f(D)} X$

$\quad$ P.I. $= \dfrac{1}{D^2 + 3D + 2} \sin e^x$

$\qquad = \dfrac{1}{(D+2)(D+1)} \sin e^x$

$\qquad = \dfrac{1}{D+2} e^{-x} \displaystyle\int e^x . \sin e^x \, dx$

$\qquad = \dfrac{1}{D+2} e^{-x} \displaystyle\int \sin t \, dt \qquad\qquad \{\because \text{ put } e^x = t, e^x dx = dt$

$\qquad = \dfrac{1}{D+2} e^{-x} (-\cos e^x) \qquad\qquad \{\because t = e^x$

$\qquad = \dfrac{-1}{D+2} (e^{-x} \cos e^x)$

$\qquad = -e^{-2x} \displaystyle\int e^{2x} (e^{-x} \cos e^x) \, dt$

$\qquad = -e^{-2x} \displaystyle\int e^x . \cos e^x \, dt \{\because \text{ put } e^x = t, \ e^x dx = dt$

$\quad$ **P.I.** $= -e^{-2x} \sin e^x$

Type II: $X = e^{ax}$

Example 8: Find the P.I. of $(D^2 + 5D + 6)y = e^x$

Solution: P.I. $= \dfrac{1}{f(D)} X$

$\quad$ P.I. $= \dfrac{1}{D^2 + 5D + 6} e^x \quad = \dfrac{e^x}{1^2 + 5(1) + 6}$

$\quad$ P.I. $= \dfrac{e^x}{12}$

Example 9: Find the P.I. of $(D+2)(D-1)^2 y = e^{-2x} + 2 \sin hx$

Solution: P.I. $= \dfrac{1}{f(D)} X$

$$P.I. = \frac{1}{(D+2)(D-1)^2}[e^{-2x} + 2\sin hx]$$

$$= \frac{1}{(D+2)(D-1)^2}[e^{-2x} + e^x - e^{-x}] \qquad \left\{ \because \sin h\theta = \frac{e^\theta - e^{-\theta}}{2} \right.$$

$$= \frac{1}{(D+2)(D-1)^2}e^{-2x} + \frac{1}{(D+2)(D-1)^2}e^x - \frac{1}{(D+2)(D-1)^2}e^{-x}$$

$$= \frac{1}{D+2}\frac{e^{-2x}}{(-2-1)^2} + \frac{1}{(D-1)^2}\frac{e^x}{(1+2)} - \frac{e^{-x}}{(-1+2)(-1-1)^2}$$

$$= \frac{1}{9}\frac{x}{1}e^{-2x} + \frac{1}{3}\frac{x^2}{2!}\cdot e^x - \frac{e^{-x}}{4}$$

$$P.I. = \frac{x}{9}e^{-2x} + \frac{x^2}{6}e^x - \frac{1}{4}e^{-x}$$

Type III: X = sin(ax + b) or cos(ax + b)

Example 10: Find the P.I. of $\dfrac{d^3y}{dx^3} + 4\dfrac{dy}{dx} = \sin 2x$

Solution: $P.I. = \dfrac{1}{f(D)}X$, Given equation in symbolic form is $(D^3 + 4D)y = \sin 2x$

$$P.I. = \frac{1}{D^3 + 4D}\sin 2x$$

$$= \frac{x}{3D^2 + 4}\sin 2x \left\{ \begin{array}{l} \because \text{puting } D^2 = -(2)^2 = -4 \\ D.D^2 = -4D + 4D = 0 \end{array} \right.$$

$$= \frac{x.\sin 2x}{3(-1)(2)^2 + 4} = \frac{x.\sin 2x}{-8}$$

$$\therefore \quad P.I. = \frac{-x}{8}\sin 2x$$

Example 11: Find the P.I. of $(D^3 + 1)y = \cos(2x - 1)$

Solution: $P.I. = \dfrac{1}{f(D)}X$

$$P.I. = \frac{1}{D^3 + 1}\cos(2x - 1)$$

$$= \frac{1}{D.D^2 + 1}\cos(2x - 1)$$

$$= \frac{1}{D(-4) + 1}\cos(2x - 1)\{\because \text{put } D^2 = (-2)^2 = -4$$

$$= \frac{1}{1 - 4D}\cos(2x - 1)$$

$$= \frac{1}{1 - 4D}\cdot\frac{1 + 4D}{1 + 4D}\cos(2x - 1) \qquad \{\because \text{Rationalised } 1 - 4D$$

$$= \frac{1 + 4D}{1 - 16D^2} \cos(2x - 1)$$

$$= (1 + 4D)\frac{1}{1 - 16(-4)} \cos(2x - 1)$$

$$= (1 + 4D)\frac{1}{65} \cos(2x - 1)$$

$$= \frac{1}{65}[\cos(2x - 1) + 4D\cos(2x - 1)]$$

P. I. $= \dfrac{1}{65}[\cos(2x - 1) - 8\sin(2x - 1)]$

Type IV: X = x^m , m is a positive integer

Example 12: Find the P. I. of $\dfrac{d^2y}{dx^2} + \dfrac{dy}{dx} = x^2 + 2x + 4$

Solution: P. I. $= \dfrac{1}{f(D)}X$

Given equation in symbolic form is $(D^2 + D)y = x^2 + 2x + 4$

P. I. $= \dfrac{1}{D^2 + D}(x^2 + 2x + 4)$

$= \dfrac{1}{D(D + 1)}(x^2 + 2x + 4)$

$= \dfrac{1}{D}(1 + D)^{-1}(x^2 + 2x + 4)$

$= \dfrac{1}{D}(1 - D + D^2 - D^3 + \cdots)(x^2 + 2x + 4)$

$= \dfrac{1}{D}[x^2 + 2x + 4 - D(x^2 + 2x + 4) + D^2(x^2 + 2x + 4)]$

$= \dfrac{1}{D}[x^2 + 2x + 4 - (2x + 2) + 2]$

$= \dfrac{1}{D}[x^2 + 4] = \displaystyle\int (x^2 + 4)dx = \int x^2 dx + 4\int 1 dx$

P. I. $= \dfrac{x^2}{3} + 4x$

Type V: X = e^{ax}.V

Example 13: Find P. I. of $(D^2 - 2D + 4)y = e^x . \cos x$

Solution: P. I. $= \dfrac{1}{f(D)}X$

P. I. $= \dfrac{1}{D^2 - 2D + 4}e^x \cos x$

$$= e^x \frac{1}{(D+1)^2 - 2(D+1) + 4} \cos x$$

$$= e^x \frac{1}{D^2 + 2D + 1 - 2D - 2 + 4} \cos x$$

$$= e^x \frac{1}{D^2 + 3} \cos x$$

$$= e^x \frac{\cos x}{-(1)^2 + 3}$$

$$\text{P. I.} = \frac{1}{2} e^x \cos x$$

6   Rules for finding the Complete Solution (C. S.)

Working procedure to solve equation

Step I: To find the complementary function (C. F.)

Step II: To find the particular integral (P. I.)

Step III: To find the complete solution (C. S.) or general solution (G. S.)

$$\text{C. S. is } \quad y = \text{ C. F.} + \text{P. I.}$$

Examples on Complete Solution (C. S.)

6. i   Examples on:

Type I:

$$X = e^{e^x}, \sin e^x, \cos e^x, \sec x, \csc x, \tan x, \cot x, \log x, \frac{1}{x}, \frac{1}{1+x^2}$$

i) $f(D) = D - a$

$$\therefore \frac{1}{D-a} X = e^{ax} \int e^{-ax} X \, dx$$

   ii) $f(D) = D + a$

$$\therefore \frac{1}{D+a} X = e^{-ax} \int e^{ax} X \, dx$$

Example 14: Solve $\dfrac{d^2 y}{dx^2} + 3 \dfrac{dy}{dx} + 2y = \sin(e^x)$

Solution: Given equation in symbolic form is $(D^2 + 3D + 2)y = \sin(e^x)$

   C. S. = C. F. + P. I.

i) To find C. F.

   Its A. E. is $D^2 + 3D + 2 = 0$

$\therefore (D+1)(D+2) = 0$; $\therefore D = -1, -2$

Thus C. F. $= C_1 e^{-x} + C_2 e^{-2x}$

ii) To find P. I.

$$P.I. = \frac{1}{D^2 + 3D + 2} \sin(e^x)$$

$$= \frac{1}{(D + 2)(D + 1)} \sin(e^x)$$

$$= \frac{1}{(D + 2)} e^{-x} \int e^x \sin(e^x) \, dx \left\{ \because \int \sin\{f(x)\}. f'(x) dx = -\cos\{f(x)\} \right\}$$

$$= \frac{1}{D + 2} [e^{-x}(-\cos e^x)]$$

$$= e^{-2x} \int e^{2x} e^{-x}(-\cos e^x) \, dx$$

$$= -e^{-2x} \int e^x. \cos e^x \, dx \left\{ \because \int \cos [f(x)]. f(x) dx = \sin\{f(x)\} \right\}$$

$$= -e^{-2x} \sin e^x$$

iii) Hence the C. S. is $y = C_1 e^{-x} + C_2 e^{-2x} - e^{-2x} \sin(e^x)$

Example 15: Solve $(D^2 + a^2)y = \tan(ax)$

Solution: Given the D. E. is, $(D^2 + a^2)y = \tan(ax)$

  C. S. = C. F. + P. I.

i) To find C. F.

Its A. F. is $D^2 + a^2 = 0$; $\therefore D^2 = -a^2$; $\therefore D = \pm ai$

Thus C. F. = $C_1 \cos ax + C_2 \sin ax$

ii) To find P. I.

$$P. I. = \frac{1}{D^2 + a^2} \tan(ax)$$

$$= \frac{1}{D^2 - i^2 a^2} \tan(ax) \{\because \text{Note } i^2 = -1$$

$$= \frac{1}{(D - ia)(D + ia)} \tan(ax)$$

$$= \frac{1}{2ia} \left[\frac{1}{D - ia} - \frac{1}{D + ia} \right] \tan(ax) \qquad \qquad ... \text{Note} \quad \{\because \text{partial fraction}$$

$$= \frac{1}{2ia} \left[\frac{1}{D - ia} \tan(ax) - \frac{1}{D + ia} \tan(ax) \right]$$

$$= \frac{1}{2ia} \left[e^{iax} \int e^{-iax} \tan(ax) \, dx - e^{-iax} \int e^{iax} \tan(ax) \, dx \right]$$

$$= \frac{1}{2ia} \left[e^{iax} \int (\cos ax - i \sin ax). \tan ax \, dx - e^{-iax} \int (\cos ax + i \sin ax) \tan ax \, dx \right]$$

$$= \frac{1}{2ia}\left[e^{iax} \int \left(\sin ax - \frac{i\sin^2 ax}{\cos ax} \right) dx - e^{-iax} \int \left(\sin ax + \frac{i\sin^2 ax}{\cos ax} \right) dx \right]$$

$$= \frac{1}{2ia}\left[e^{iax} \int (\sin ax - \frac{i(1 - \cos^2 ax)}{\cos ax} dx - e^{-iax} \int \left(\sin ax + \frac{i(1 - \cos^2 ax)}{\cos ax} \right) dx \right]$$

$$= \frac{1}{2ia}\left[\left(e^{iax} - e^{-iax} \right) \int \sin ax \, dx - i\left(e^{iax} + e^{-iax} \right) \int \left(\frac{1}{\cos ax} - \cos ax \right) dx \right]$$

$$= \frac{1}{2ia}\left\{ 2i\sin ax \left(\frac{-\cos ax}{a} \right) - 2i\cos ax \left[\left(\frac{\log (\sec ax + \tan ax)}{a} \right) - \frac{\sin ax}{a} \right] \right\}$$

$$= \frac{1}{2ia}\left[\frac{-2i\sin ax \cos ax}{a} - \frac{2i\cos ax}{a}\log(\sec ax + \tan ax) + \frac{2i\sin ax.\cos ax}{a} \right]$$

$$= \frac{-1}{2ia}\frac{2i}{a}\cos ax . \log (\sec ax + \tan ax)$$

$$= \frac{-\cos ax}{a^2} \log (\sec ax + \tan ax)$$

iii) **Hence C. S. is** $y = C_1\cos ax + C_2\sin ax - \dfrac{\cos ax}{a^2}\log (\sec ax + \tan ax)$

Example 16: Solve: $\dfrac{d^2y}{dx^2} + a^2y = \sec ax$

Solution: In symbolic form of given DE is, $(D^2 + a^2)y = \sec ax$

C. S. = C. F. + P. I.

i) **To find C. F.**

Its A. E. is $D^2 + a^2 = 0$; $D^2 = -a^2$

$\therefore D = \pm\, ia$

$\therefore$ C. F. $= C_1 \cos ax + C_2 \sin ax$

ii) **To find P. I.**

$$\text{P. I.} = \frac{1}{D^2 + a^2}\sec ax$$

$$= \frac{1}{D^2 - i^2a^2}\sec ax \qquad\qquad \{\because i^2 = -1$$

$$= \frac{1}{(D - ia)(D + ia)}\sec ax \qquad \{\because \text{By factorising } a^2 - b^2 = (a - b)(a + b)$$

$$= \frac{1}{2ia}\left[\frac{1}{D - ia} - \frac{1}{D + ia} \right]\sec ax \qquad\qquad \{\because \text{By partial function}$$

$$= \frac{1}{2ia}\left[\frac{1}{D - ia}\sec ax - \frac{1}{D + ia}\sec ax \right]$$

$$= \frac{1}{2ia}\left[e^{iax} \int e^{-ax}\sec ax \, dx - e^{-iax} \int e^{iax}\sec ax \, dx \right]$$

$$= \frac{1}{2ia}\left[e^{iax}\int (\cos ax - i\sin ax)\sec ax\, dx - e^{-iax}\int (\cos ax + i\sin ax)\sec ax\, dx\right]$$

$$= \frac{1}{2ia}\left[e^{iax}\int (1 - i\tan ax)\, dx - e^{-iax}\int (1 + i\tan ax)\, dx\right]$$

$$= \frac{1}{2ia}\left[\left(e^{iax} - e^{-iax}\right)\int 1\, dx - i\left(e^{iax} + e^{-iax}\right)\int \tan ax\, dx\right]$$

$$= \frac{1}{2ia}\left[2i\sin ax.(x) - i\, 2\cos ax.\frac{\log|\sec ax|}{a}\right]$$

$$= \frac{1}{2ia}\, 2i\left[x\sin ax - \cos ax.\frac{\log|\sec ax|}{a}\right]$$

$$= \frac{x}{a}\sin ax - \frac{1}{a^2}\cos ax.\log|\sec ax|$$

iii) $\therefore$ **C. S. is** $\quad y = C_1 \cos ax + C_2 \sin ax + \dfrac{x}{a}\sin ax - \dfrac{1}{a^2}\cos ax.\log|\sec ax|$

Example 17: Solve $\quad \dfrac{d^2y}{dx^2} - \dfrac{dy}{dx} - 2y = 2\log x + \dfrac{1}{x} + \dfrac{1}{x^2}$

Solution: Given D. E. is in symbolic forms as $\quad (D^2 - D - 2)y = 2\log x + \dfrac{1}{x} + \dfrac{1}{x^2}$

C. S. = C. F. + P. I.

i) To find C. F.

Its A. E. is $D^2 - D - 2 = 0$ $\therefore (D+1)(D-2) = 0$, $D = -1, 2$

$\therefore$ C. F. = $C_1 e^{-x} + C_2 e^{2x}$

ii) To find P. I.

$$P. I. = \frac{1}{D^2 - D - 2}\left[2\log x + \frac{1}{x} + \frac{1}{x^2}\right]$$

$$= \frac{1}{(D-2)(D+1)}\left(2\log x + \frac{1}{x} + \frac{1}{x^2}\right)$$

$$= \frac{1}{D-2}e^{-x}\int e^x\left(2\log x + \frac{1}{x} + \frac{1}{x^2}\right) dx$$

$$= \frac{1}{D-2}e^{-x}\int e^x\left(2\log x - \frac{1}{x} + \frac{2}{x} + \frac{1}{x^2}\right) dx \quad \{\because \text{Note down}$$

$$= \frac{1}{D-2}e^{-x}.e^x\left(2\log x - \frac{1}{x}\right)\left\{\because \int e^x[f(x) + f'(x)]\, dx = e^x\ f(x)\right.$$

$$= \frac{1}{D-2}\left(2\log x - \frac{1}{x}\right)$$

$$= e^{2x}\int e^{-2x}\left(2\log x - \frac{1}{x}\right) dx$$

$$= e^{2x}\left[2\int e^{-2x}\log x\, dx - \int e^{-2x}\frac{1}{x}\, dx\right] \quad \{\because \text{use LIATE rule}$$

$$= e^{2x} \left[2 \left(\log x \frac{e^{-2x}}{-2} - \int \frac{1}{x} \frac{e^{-2x}}{-2} \, dx \right) - \int e^{-2x} \cdot \frac{1}{x} \, dx \right]$$

$$= e^{2x} \left[-\log x \cdot e^{-2x} + \int e^{-2x} \frac{1}{x} \, dx - \int e^{-2x} \frac{1}{x} \, dx \right]$$

$$= -e^{2x \cdot} e^{-2x} . \log x$$

$$= -\log x$$

iii) $\therefore$ C.S. is $y = C_1 e^{-x} + C_2 e^{2x} - \log x$

Example 18: Solve $(D^2 + 5D + 6)y = e^{-2x} \sec^2 x \, (1 + 2 \tan x)$

Solution: Given D. E. is, $(D^2 + 5D + 6)y = e^{-2x} \sec^2 x \, (1 + 2 \tan x)$

   C.S. = C.F. + P.I.

i) To find C. F.

   Its A. E. is $D^2 + 5D + 6 = 0$ $\therefore (D + 2)(D + 3) = 0$

$\therefore D = -2, \; -3$

$\therefore$ C.F. = $C_1 e^{-2x} + C_2 e^{-3x}$

ii) To find P. I.

$$P.I. = \frac{1}{D^2 + 5D + 6} e^{-2x} \sec^2 x \, (1 + 2 \tan x)$$

$$= \frac{1}{(D + 2)(D + 3)} e^{-2x} \sec^2 x \, (1 + 2 \tan x)$$

$$= \frac{1}{D + 2} e^{-3x} \int e^{3x} \, e^{-2x} \sec^2 x \, (1 + 2 \tan x) \, dx$$

$$= \frac{1}{D + 2} e^{-3x} \int e^{x} (\sec^2 x + 2\sec^2 x \tan x) \, dx \left\{ \because \int e^{x} [f(x) + f'(x)] \, dx \right.$$

$$\left. = e^{x} \; f(x) \right.$$

$$= \frac{1}{D + 2} e^{-3x} e^{x} \sec^2 x$$

$$= e^{-2x} \int e^{2x} \, e^{-3x} \, e^{x} \sec^2 x \, dx$$

$$= e^{-2x} \int \sec^2 x \, dx \left\{ \because \int \sec^2 \theta \, d\theta = \tan \theta \right.$$

$$= e^{-2x} \tan x$$

iii) $\therefore$ C.S. is $y = C_1 e^{-2x} + C_2 e^{-3x} + e^{-2x} \tan x$

Example 19: Solve $(D^2 + D)y = (1 + e^x)^{-1}$

Solution: Given D. E. is, $(D^2 + D)y = (1 + e^x)^{-1}$

$$\text{i. e. } (D^2 + D)y = \frac{1}{1 + e^x}$$

$$C.S. = C.F. + P.I.$$

i) To find C.F.

Its A. E. is $D^2 + D = 0$; $\therefore$ $D(D + 1) = 0$; $\therefore$ $D = 0, -1$

$\therefore C.F. = C_1 e^{0x} + C_2 e^{-x} = C_1 + C_2 e^{-x}$

ii) To find P. I.

$$P.I. = \frac{1}{D^2 + D}\frac{1}{1 + e^x}$$

$$= \frac{1}{D(D + 1)}\frac{1}{1 + e^x}$$

$$= \frac{1}{D}e^{-x}\int e^x \frac{1}{1 + e^x}\ dx$$

$$= \frac{1}{D}e^{-x}\log(1 + e^x)\left\{\because \int \frac{f'(x)}{f(x)}dx = \log f(x)\right.$$

$$= \int e^{-x}\log(1 + e^x)\ dx$$

$$= \log(1 + e^x).\frac{e^{-x}}{-1} - \int \frac{1}{1 + e^x}.e^x(-e^{-x})\ dx \qquad \{\because \text{ By LIATE rule}$$

$$= -e^{-x}\log(1 + e^x) + \int \frac{1}{1 + e^x}\ dx$$

$$= -e^{-x}\log(1 + e^x) + \int \frac{1}{e^x(e^{-x} + 1)}\ dx \ ... \text{Note}$$

$$= -e^{-x}\log(1 + e^x) - \int \frac{-e^{-x}}{1 + e^{-x}}\ dx\left\{\because \int \frac{f'(x)}{f(x)}\ dx = \log f(x)\right.$$

$$= -e^{-x}\log(1 + e^x) - \log(1 + e^{-x})$$

iii) $\therefore$ **C. S. is** $y = C_1 + C_2 e^{-x} - e^{-x}\log(1 + e^x) - \log(1 + e^{-x})$

Example 20: Solve $\dfrac{d^2y}{dx^2} + 3\dfrac{dy}{dx} + 2y = e^{e^x}$

Solution: Given D. E. in symbolic form as $(D^2 + 3D + 2)y = e^{e^x}$

$$C.S. = C.F. + P.I.$$

i) To find C. F.

Its A. E. is $D^2 + 3D + 2 = 0$; $\therefore$ $(D + 1)(D + 2) = 0$; $\therefore$ $D = -1, -2$

$\therefore C.F. = C_1 e^{-x} + C_2 e^{-2x}$

ii) To find P. I.

$$P.I. = \frac{1}{D^2 + 3D + 2}e^{e^x}$$

$$= \frac{1}{(D + 1)(D + 2)}e^{e^x}$$

$$= \left[\frac{1}{D+1} - \frac{1}{D+2}\right]e^{e^x}\{\because \text{ By Partial fraction}$$

$$= \frac{1}{D+1}e^{e^x} - \frac{1}{D+2}e^{e^x}$$

$$= e^{-x}\int e^x e^{e^x}\,dx - e^{-2x}\int e^{2x}e^{e^x}\,dx$$

$$= e^{-x}\int e^t\,dt - e^{-2x}\int te^t\,dt\begin{cases} \because \text{ put } e^x = t; \\ e^x\,dx = dt \end{cases}$$

$$= e^{-x}e^t - e^{-2x}(te^t - (1)e^t) \qquad\qquad \{\because \text{ By LIATE rule}$$

$$= e^{-x}e^{e^x} - e^{-2x}(e^x . e^{e^x} - e^{e^x})$$

$$= e^{-x}e^{e^x} - e^{-x}e^{e^x} + e^{-2x}e^{e^x}$$

$$= e^{-2x}. e^{e^x}$$

iii) C. S. is $y = C_1 e^{-x} + C_2 e^{-2x} + e^{-2x}. e^{e^x}$

---

6. ii   Examples on:

Type II: $X = e^{ax}$

i) $\dfrac{1}{f(D)} e^{ax} = \dfrac{e^{ax}}{f(a)}$,   provided $f(a) \ne 0$

ii) $\dfrac{1}{(D-a)^r} e^{ax} = \dfrac{x^r}{r!}e^{ax}$

Example 21: Solve $\left(D^4 - 6D^3 + 11D^2 - 6D\right)y = e^{-3t}$, where $D = \dfrac{d}{dt}$

Solution: Given D. E. is, $(D^4 - 6D^3 + 11D^2 - 6D)y = e^{-3t}$

   C. S. = C. F. + P. I.

i) To find C. F.

Its A. E. is $D^4 - 6D^3 + 11D^2 - 6D = 0$

i. e. $D(D^3 - 6D^2 + 11D - 6) = 0$

i. e. $D(D-1)(D-2)(D-3) = 0$

$\therefore D = 0, \ 1, \ 2, \ 3$

$\therefore$ C. F. $= C_1 + C_2 e^t + C_3 e^{2t} + C_4 e^{3t}$

ii) To find P. I.

$$\text{P. I.} = \frac{1}{D(D-1)(D-2)(D-3)}e^{-3t}$$

$$= \frac{1. e^{-3t}}{(-3)(-3-1)(-3-2)(-3-3)} = \frac{e^{-3t}}{(-3)(-4)(-5)(-6)}$$

$$= \frac{1}{360}e^{-3t}$$

iii) $\therefore$ C. S. is $\quad y = C_1 + C_2 e^t + C_3 e^{2t} + C_4 e^{3t} + \dfrac{e^{-3t}}{360}$

Example 22: Solve $(D^3+1)y = 3 + e^{-x}+5e^{2x}$

Solution: Given D. E. is, $\quad (D^3 + 1)y = 3 + e^{-x} + 5e^{2x}$

$\quad$ C. S. = C. F. + P. I.

i) To find C. F.

Its A. E. is $D^3 + 1 = 0$

$(D + 1)(D^2 - D + 1) = 0 \qquad \{\because a^2 + b^3 = (a + b)(a^2 - ab + b^2)$

$\therefore D = -1, \quad \dfrac{1 \pm \sqrt{(-1)^2 - 4(1)(1)}}{2(1)} \left\{ \because ax^2 + bx + c \Rightarrow x = \dfrac{-b \pm \sqrt{b^2 - 4ac}}{2a} \right.$

$D = -1, \quad \dfrac{1 \pm \sqrt{(1 - 4)}}{2} \qquad \text{i.e } D = -1, \ \dfrac{1}{2} \pm i\dfrac{\sqrt{3}}{2}$

$\therefore$ C. F. $= C_1 e^{-x} + e^{\frac{x}{2}}\left(C_2 \cos \dfrac{\sqrt{3}}{2}x + C_3 \sin \dfrac{\sqrt{3}}{2}x \right)$

ii) To find P. I.

$$\text{P. I.} = \frac{1}{D^3 + 1}(3 + e^{-x} + 5e^{2x})$$

$$= 3\frac{1}{D^3 + 1}e^{0x} + \frac{1}{D^3 + 1}e^{-x} + 5\frac{1}{D^3 + 1}e^{2x}$$

$$= 3\frac{e^{0x}}{0 + 1} + \frac{x}{3D^2}e^{-x} + 5\frac{e^{2x}}{2^3 + 1}$$

$$= 3 + \frac{x e^{-x}}{3(-1)^2} + \frac{5e^{2x}}{9}$$

$$= 3 + \frac{x e^{-x}}{3(-1)^2} + \frac{5}{9}e^{2x}$$

$$= 3 + \frac{x}{3}e^{-x} + \frac{5}{9}e^{2x}$$

iii) $\therefore$ C. S. is $\quad y = C_1 e^{-x} + e^{\frac{x}{2}}\left(C_2 \cos\dfrac{\sqrt{3}}{2}x + C_3 \sin \dfrac{\sqrt{3}}{2}x \right) + 3 + \dfrac{x}{3}e^{-x} + \dfrac{5}{9}e^{2x}$

Example 23: Solve $\left(D^2 - D - 6\right) y = e^x \cosh 2x$

Solution: Given D. E. is, $\quad (D^2 - D - 6) y = e^x \cosh 2x$

$\quad$ C. S. = C. F. + P. I.

i) To find C. F.

Its A. E. is $D^2 - D - 6 = 0$; $(D - 3)(D + 2) = 0$

$\therefore D = 3, -2$

$\therefore \text{C. F.} = C_1 e^{-2x} + C_2 e^{3x}$

ii) To find P. I.

$$\text{P. I.} = \frac{1}{(D - 3)(D + 2)} e^x \cdot \cosh 2x$$

$$= \frac{1}{(D - 3)(D + 2)} e^x \cdot \left(\frac{e^{2x} + e^{-2x}}{2}\right) \left\{ \because \cosh \theta = \frac{e^\theta + e^{-\theta}}{2} \right.$$

$$= \frac{1}{(D - 3)(D + 2)} \frac{1}{2}(e^{3x} + e^{-x})$$

$$= \frac{1}{2}\left[\frac{1}{(D - 3)(D + 2)} e^{3x} + \frac{1}{(D - 3)(D + 2)} e^{-x} \right]$$

$$= \frac{1}{2}\left[\frac{1}{(D - 3)} \frac{e^{3x}}{(3 + 2)} + \frac{e^{-x}}{(-1 - 3)(-1 + 2)} \right]$$

$$= \frac{1}{2}\left[\frac{1}{5}\frac{x}{1} e^{3x} + \frac{e^{-x}}{-4} \right]$$

$$= \frac{1}{10} x\, e^{3x} - \frac{1}{8} e^{-x}$$

iii) $\therefore$ **C. S. is** $y = C_1 e^{-2x} + C_2 e^{3x} + \dfrac{1}{10} x\, e^{3x} - \dfrac{1}{8} e^{-x}$

Example 24: Solve $(D^2 + 13D + 36)y = e^{-4x} + \sinh x$

Solution: Given D. E. is, $(D^2 + 13D + 36)y = e^{-4x} + \sinh x$

  C. S. = C. F. + P. I.

i) To find C. F.

Its A. E. is $D^2 + 13D + 36 = 0$; $(D + 4)(D + 9) = 0$

$\therefore D = -4, -9$

$\therefore \text{C. F.} = C_1 e^{-4x} + C_2 e^{-9x}$

ii) To find P. I.

$$\text{P. I.} = \frac{1}{D^2 + 13D + 36} (e^{-4x} + \sinh x)$$

$$= \frac{1}{(D + 4)(D + 9)} \left(e^{-4x} + \frac{e^x - e^{-x}}{2}\right)$$

$$= \frac{1}{(D + 4)(D + 9)} e^{-4x} + \frac{1}{2}\frac{1}{(D + 4)(D + 9)} e^x - \frac{1}{2}\frac{1}{(D + 4)(D + 9)} e^{-x}$$

$$= \frac{1}{D + 4}\frac{e^{-4x}}{(-4 + 9)} + \frac{1}{2}\frac{e^x}{(1 + 4)(1 + 9)} - \frac{1}{2}\frac{e^{-x}}{(-1 + 4)(-1 + 9)}$$

$$= \frac{x}{1} \frac{e^{-4x}}{5} + \frac{1}{2} \frac{e^x}{50} - \frac{1}{2} \frac{e^{-x}}{24}$$

$$= \frac{x}{5} e^{-4x} + \frac{e^x}{100} - \frac{e^{-x}}{48}$$

iii) ∴ C.S. is $y = C_1 e^{-4x} + C_2 e^{-9x} + \frac{x}{5} e^{-4x} + \frac{e^x}{100} - \frac{e^{-x}}{48}$

Example 25: Solve $\dfrac{d^2y}{dx^2} + \dfrac{dy}{dx} + y = (1 - e^x)^2$

Solution: Given equation in symbolic form as $(D^2 + D + 1)y = (1 - e^x)^2$

C.S. = C.F. + P.I.

i) To find C.F.

Its A.E. is $D^2 + D + 1 = 0$

$$\therefore D = \frac{-1 \pm \sqrt{(-1)^2 - 4(1)(1)}}{2(1)} = \frac{-1}{2} \pm i \frac{\sqrt{3}}{2}$$

$$\therefore C.F. = e^{-\frac{x}{2}} \left(C_1 \cos \frac{\sqrt{3}}{2} x + C_2 \sin \frac{\sqrt{3}}{2} x \right)$$

ii) To find P.I.

$$P.I. = \frac{1}{D^2 + D + 1} (1 - e^x)^2$$

$$= \frac{1}{D^2 + D + 1} (1 - 2e^x + e^{2x})$$

$$= \frac{1}{D^2 + D + 1} e^{0x} - 2 \frac{1}{D^2 + D + 1} e^x + \frac{1}{D^2 + D + 1} e^{2x}$$

$$= \frac{1}{0 + 0 + 1} - 2 \frac{e^x}{1 + 1 + 1} + \frac{e^{2x}}{2^2 + 2 + 1}$$

$$= 1 - \frac{2}{3} e^x + \frac{1}{7} e^{2x}$$

iii) ∴ C.S. is $y = e^{-\frac{x}{2}} \left(c_1 \cos \frac{\sqrt{3}}{2} x + c_2 \sin \frac{\sqrt{3}}{2} x \right) + 1 - \frac{2. e^x}{3} + \frac{e^{2x}}{7}$

---

6. iii   Examples on:

Type III: X = sin(ax + b) or cos(ax + b)

i) $\dfrac{1}{f(D^2)} \sin(ax + b) = \dfrac{\sin(ax + b)}{f(-1. a^2)}$,   provided $f(-1. a^2) \neq 0$

ii) $\dfrac{1}{f(D^2)} \cos(ax + b) = \dfrac{\cos(ax + b)}{f(-1. a^2)}$,   provided $f(-1. a^2) \neq 0$

iii) $\dfrac{1}{(D^2+a^2)^r}\sin(ax+b) = \left(\dfrac{-x}{2a}\right)^r \dfrac{1}{r!}\sin\left(ax+b+\dfrac{r\pi}{2}\right)$

iv) $\dfrac{1}{(D^2+a^2)^r}\cos(ax+b) = \left(\dfrac{-x}{2a}\right)^r \dfrac{1}{r!}\cos\left(ax+b+\dfrac{r\pi}{2}\right)$

Example 26: **Solve** $(D^3 + 1)y = \sin(2x + 3)$

Solution: Given D. E. is, $(D^3 + 1)y = \sin(2x + 3)$

    C. S. = C. F. + P. I.

i) To find C. F.

Its A. E. is $D^3 + 1 = 0$; $(D + 1)(D^2 - D + 1) = 0$

$\therefore D = -1, \quad \dfrac{1 \pm \sqrt{1 - 4}}{2}$; $D = -1, \quad \dfrac{1}{2} \pm i\dfrac{\sqrt{3}}{2}$

$\therefore$ C. F. $= C_1 e^{-x} + e^{\frac{x}{2}}\left(C_2\cos\dfrac{\sqrt{3}}{2}x + C_3\sin\dfrac{\sqrt{3}}{2}x\right)$

ii) To find P. I.

P. I. $= \dfrac{1}{D^3 + 1}\sin(2x + 3)$

$= \dfrac{1}{D.D^2 + 1}\sin(2x + 3)$

$= \dfrac{1}{D(-1)(2)^2 + 1}\sin(2x + 3)$

$= \dfrac{1}{-4D + 1}\sin(2x + 3)$

$= \dfrac{1 + 4D}{1 - 16D^2}\sin(2x + 3)$ $\{\because$ Rationalize D^r by$(1 + 4D)$

$= (1 + 4D)\dfrac{\sin(2x + 3)}{1 - 16(-1)(2)^2}$

$= (1 + 4D)\dfrac{\sin(2x + 3)}{1 + 64}$

$= \dfrac{1}{65}[\sin(2x + 3) + 4D\sin(2x + 3)]$

$= \dfrac{1}{65}[\sin(2x + 3) + 8\cos(2x + 3)]$

iii) $\therefore$ **C. S. is,**

$y = C_1 e^{-x} + e^{\frac{x}{2}}\left(C_2\cos\dfrac{\sqrt{3}}{2}x + C_3\sin\dfrac{\sqrt{3}}{2}x\right) + \dfrac{1}{65}[\sin(2x + 3) + 8\cos(2x + 3)]$

Exmple 27: **Solve** $(D^2 - 2D + 5)\,y = \sin 3x$

Solution: Given D. E. is, $(D^2 - 2D + 5) y = \sin 3x$

  C. S. = C. F. + P. I.

i) To find C. F.

Its A. E. is $D^2 - 2D + 5 = 0$;

$$D = \frac{2 \pm \sqrt{4 - 20}}{2} = \frac{2 \pm 4i}{2} \quad ; \quad \therefore D = 1 \pm 2i$$

$\therefore$ C. F. $= e^x (C_1 \cos 2x + C_2 \sin 2x)$

ii) To find P. I.

$$
\begin{aligned}
\text{P. I.} \; &= \frac{1}{D^2 - 2D + 5} \sin 3x \\
&= \frac{1}{(-1)(3)^2 - 2D + 5} \sin 3x \\
&= \frac{1}{-2D - 4} \sin 3x \\
&= -\frac{1}{2} \frac{1}{(D + 2)} \sin 3x \\
&= -\frac{1}{2} \frac{(D - 2)}{D^2 - 4} \sin 3x \qquad \{\because \text{Rationlize } D^r \text{ by } D - 2 \\
&= \frac{-1}{2} (D - 2) \frac{\sin 3x}{(-1)(3)^2 - 4} \\
&= \frac{-1}{2} \frac{1}{-13} (D \sin 3x - 2 \sin 3x) \\
&= \frac{1}{26} (3 \cos 3x - 2 \sin 3x)
\end{aligned}
$$

iii) $\therefore$ **C. S. is** $\quad y = e^x(C_1 \cos 2x + C_2 \sin 2x) + \dfrac{1}{26} (3 \cos 3x - 2 \sin 3x)$

Example 28: Solve $\dfrac{d^3 y}{dx^3} + y = \sin 3x - \cos^2 \dfrac{x}{2}$

Solution: Given D. E. in symbolic form as $(D^2 + 1)y = \sin 3x - \cos^2 \dfrac{x}{2}$

  C. S. = C. F. + P. I.

i) To find C. F.

  Its A. E. is $D^3 + 1 = 0$; $(D + 1)(D^2 - D + 1) = 0$

$\therefore D = -1, \quad \dfrac{1 \pm \sqrt{1 - 4}}{2}$; $\quad D = -1, \quad \dfrac{1}{2} \pm i \dfrac{\sqrt{3}}{2}$

$\therefore$ C. F. $= C_1 e^{-x} + e^{\frac{x}{2}} \left(C_2 \cos \dfrac{\sqrt{3}}{2} x + C_3 \sin \dfrac{\sqrt{3}}{2} x \right)$

ii) To find P. I.

P.I. $= \dfrac{1}{D^3 + 1}\left(\sin 3x - \cos^2\dfrac{x}{2}\right)$

$= \dfrac{1}{D^3 + 1}\sin 3x - \dfrac{1}{D^3 + 1}\cos^2\dfrac{x}{2}$

$= \dfrac{1}{D^2.D + 1}\sin 3x - \dfrac{1}{D^3 + 1}\left(\dfrac{1 + \cos x}{2}\right)\left\{\because \cos^2\theta = \dfrac{1 + \cos 2\theta}{2}\right\}$

$= \dfrac{1}{(-1)(3)^2 D + 1}\sin 3x - \dfrac{1}{2}\left[\dfrac{1}{D^3 + 1}\ 1 + \dfrac{1}{D^3 + 1}\cos x\right]$

$= \dfrac{1}{-9D + 1}\sin 3x - \dfrac{1}{2}\left[\dfrac{1}{D^3 + 1}e^{0x} + \dfrac{1}{D^2.D + 1}\cos x\right]$

$= \dfrac{1}{1 - 9D}\dfrac{(1 + 9D)}{(1 + 9D)}\sin 3x - \dfrac{1}{2}\left[\dfrac{1}{0 + 1} + \dfrac{1}{(-1)(1)D + 1}\cos x\right]$

$= (1 + 9D)\dfrac{1}{1 - 81D^2}\sin 3x - \dfrac{1}{2}\left[1 + \dfrac{1}{-D + 1}\cos x\right]$

$= (1 + 9D)\dfrac{\sin 3x}{1 - 81(-1)(3)^2} - \dfrac{1}{2}\left[1 + \dfrac{1 + D}{1 - D^2}\cos x\right]$

$= \dfrac{1}{730}(\sin 3x + 9D \sin 3x) - \dfrac{1}{2} - \dfrac{1}{2}\dfrac{(\cos x + D\ \cos x)}{1 - (-1)(1)^2}$

$= \dfrac{1}{730}(\sin 3x + 27 \cos 3x) - \dfrac{1}{2} - \dfrac{1}{4}(\cos x - \sin x)$

iii) $\therefore$ **C. S. is** $\mathbf{y = C_1 e^{-x} + e^{\frac{x}{2}}\left(C_2 \cos\dfrac{\sqrt{3}}{2}x + C_3 \sin\dfrac{\sqrt{3}}{2}x\right)}$

$\qquad\qquad + \dfrac{1}{730}(\sin 3x + 27 \cos 3x) - \dfrac{1}{2} - \dfrac{1}{4}(\cos x - \sin x)$

Example 29: **Solve** $(D^2 - 4D + 4)y = 4(e^{2x} - \cos 2x)$

Solution: Given D. E. is $(D^2 - 4D + 4)y = 4(e^{2x} - \cos 2x)$

$\quad$ C. S. = C. F. + P. I.

i) To find C. F.

$\quad$ Its A. E. is $D^2 - 4D + 4 = 0$; $(D - 2)^2 = 0$

$\therefore D = 2, \ 2$

$\therefore$ C. F. $= (C_1 + C_2 x)e^{2x}$

ii) To find P. I.

P.I. $= \dfrac{1}{D^2 - 4D + 4}4(e^{2x} - \cos 2x)$

$= 4\left[\dfrac{1}{D^2 - 4D + 4}e^{2x} - \dfrac{1}{D^2 - 4D + 4}\cos 2x\right]$

$$= 4\left[\frac{1}{(D-2)^2}e^{2x} - \frac{1}{D^2-4D+4}\cos 2x\right]$$

$$= 4\left[\frac{x^2}{2!}e^{2x} - \frac{1}{(-1)(2)^2-4D+4}\cos 2x\right]$$

$$= 4\left[\frac{x^2}{2}e^{2x} + \frac{1}{4}\frac{1}{D}\cos 2x\right] \qquad = 2x^2e^{2x} + \int \cos 2x\ dx \qquad = 2x^2e^{2x} + \frac{\sin 2x}{2}$$

iii) C. S. is $y = (C_1+C_2x)e^{2x} + 2x^2e^{2x} + \dfrac{\sin 2x}{2}$

Example 30: Solve $(D^4-m^4)y = \sin mx$

Solution: Given D. E. is $(D^4-m^4)y = \sin mx$

   C. S. = C. F. + P. I.

i) To find C. F.

Its A. E. is $D^4-m^4 = 0$; $(D^2)^2-(m^2)^2 = 0$

$\therefore (D^2-m^2)(D^2+m^2) = 0$

i. e. $D^2 = m^2,\ -m^2$ i. e. $D = \pm m,\ \pm mi$

$\therefore D = -m,\ m,\ \pm mi$

$\therefore$ C. F. = $C_1e^{-mx} + C_2e^{mx} + C_3\cos mx + C_4\sin mx$

ii) To find P. I.

$$\text{P. I.} = \frac{1}{D^4-m^4}\sin mx$$

$$= \frac{1}{(D^2+m^2)(D^2-m^2)}\sin mx$$

$$= \frac{1}{D^2+m^2}\frac{\sin mx}{(-1)m^2-m^2}$$

$$= \frac{1}{D^2+m^2}\frac{\sin mx}{-2m^2}$$

$$= \frac{-1}{2m^2}\frac{x}{2D}\sin mx \qquad = \frac{-x}{4m^2}\int \sin mx\ dx \qquad = \frac{-x}{4m^2}\left(\frac{-\cos mx}{m}\right)$$

$$= \frac{x}{4m^3}\cos mx$$

iii) $\therefore$ **C. S. is** $y = C_1e^{-mx} + C_2e^{mx} + C_3\cos mx + C_4\sin mx + \dfrac{x}{4m^3}\cos mx$

Example 31: Solve $(D^3 + D)y = \cos x$

Solution: Given D. E. is $(D^3 + D)y = \cos x$

C. S. = C. F. + P. I.

i) To find C. F.

Its A. E. is $D^3 + D = 0$; $D(D^2 + 1) = 0$; $D = 0$ or $(D^2 + 1) = 0$

$\therefore D = 0$ or $D^2 = -1$

$\therefore D = 0, \pm i$

$\therefore$ C. F. $= C_1 + C_2 \cos x + C_3 \sin x$

ii) To find P. I.

$$P.I. = \frac{1}{D^3 + D} \cos x$$

$$= \frac{x}{3D^2 + 1} \cos x \qquad \{\because \text{ If put } D^2 = -1 \text{ then } D^r \text{ vanishes}$$

$$= \frac{x \cdot \cos x}{3(-1)(1)^2 + 1} \quad = \frac{-x}{2} \cos x$$

iii) $\therefore$ C. S. is $y = C_1 + C_2 \cos x + C_3 \sin x - \dfrac{x}{2} \cos x$

Example 32: Solve $y'' + 4y' + 4y = 3 \sin x + 4 \cos x$, $y(0) = 1$ & $y'(0) = 0$

Solution: Given D. E. in symbolic form as $(D^2 + 4D + 4)y = 3 \sin x + 4 \cos x$

   C. S. $=$ C. F. $+$ P. I.

i) To find C. F.

Its A. E. is $D^2 + 4D + 4 = 0$; $(D + 2)^2 = 0$; $\therefore D = -2, -2$

$\therefore$ C. F. $= (C_1 + C_2 x) e^{-2x}$

ii) To find P. I.

$$P.I. = \frac{1}{D^2 + 4D + 4} (3 \sin x + 4 \cos x)$$

$$= 3 \frac{1}{D^2 + 4D + 4} \sin x + 4 \frac{1}{D^2 + 4D + 4} \cos x$$

$$= 3 \frac{1}{(-1)(1)^2 + 4D + 4} \sin x + 4 \frac{1}{(-1)(1)^2 + 4D + 4} \cos x$$

$$= 3 \frac{1}{4D + 3} \sin x + 4 \frac{1}{4D + 3} \cos x$$

$$= 3 \frac{(4D - 3)}{16D^2 - 9} \sin x + 4 \frac{(4D - 3)}{16D^2 - 9} \cos x$$

$$= \frac{3 (4D - 3) \sin x}{16(-1)(1)^2 - 9} + \frac{4 (4D - 3) \cdot \cos x}{16(-1)(1)^2 - 9}$$

$$= \frac{3}{-25} (4D \sin x - 3 \sin x) + \frac{4}{-25} (4D \cos x - 3 \cos x)$$

$$= \frac{3}{-25} (4 \cos x - 3 \sin x) - \frac{4}{25} (4(-\sin x) - 3 \cos x)$$

$$= -\frac{12}{25}\cos x + \frac{9}{25}\sin x + \frac{16}{25}\sin x + \frac{12}{25}\cos x$$

$$= \frac{9+16}{25}\sin x \quad = \frac{25}{25}\sin x$$

$$= \sin x$$

iii) ∴ C.S. is $\;y = (C_1 + C_2 x)e^{-2x} + \sin x \qquad \ldots \ldots (1)$

Now, Given $y(0) = 1 \; \& y'(0) = 0$

For y(0) = 1,   When x = 0, y = 1

∴ Equation (1) becomes, $C_1 = 1$

Also, differentiate equation (1) w. r. t. 'x'

$\quad y' = (C_1 + C_2 x)e^{-2x}(-2) + e^{-2x}C_2 + \cos x$

$y' = -2e^{-2x}C_1 - 2e^{-2x}C_2 x + e^{-2x}C_2 + \cos x$

$y' = C_2 e^{-2x} + (C_1 + C_2 x)(-2e^{-2x}) + \cos x \quad \ldots \ldots (2)$

For y'(0) = 0,   When x = 0, y' = 0

∴ Equation (2) becomes, $0 = C_2 + \big(C_1(-2)\big) + 1$

$\qquad ∴ C_2 - 2C_1 = -1$

$\qquad ∴ C_2 = -1 + 2 \qquad\qquad \{∵ C_1 = 1$

$\qquad ∴ C_2 = 1$

∴ **Required solution is** $y = (1 + x)e^{-2x} + \sin x$

Example 33: Solve $\dfrac{d^4 y}{dx^4} + 10\dfrac{d^2 y}{dx^2} + 9y = 96\sin 2x . \cos x$ **with intial**

conditions $x = 0,\; y = 0,\; \dfrac{dy}{dx} = 0,\; \dfrac{d^2 y}{dx^2} = -8 \; \& \dfrac{d^3 y}{dx^3} = -18$

Solution: Given D. E. in symbolic form is $(D^4 + 10D^2 + 9)y = 96\sin 2x . \cos x$

$\quad$ C. S. = C. F. + P. I.

i) To find C. F.

Its A. E. is $D^4 + 10D^2 + 9 = 0;\quad (D^2 + 1)(D^2 + 9) = 0$

$D^2 = -1,\; -9;\quad ∴ D = \pm i,\; \pm 3i$

∴ C. F. = $C_1 \cos x + C_2 \sin x + C_3 \cos 3x + C_4 \sin 3x$

ii) To find P. I.

$\quad$ P. I. $= \dfrac{1}{D^4 + 10D^2 + 9}(96\sin 2x . \cos x)$

$\qquad = \dfrac{1}{D^4 + 10D^2 + 9}.96\dfrac{1}{2}(\sin 3x + \sin x)$

$$= 48\left[\frac{1}{D^4+10D^2+9}\sin 3x + \frac{1}{D^4+10D^2+9}\sin x\right]$$

$$= 48\left[\frac{x}{4D^3+20D}\sin 3x + \frac{x}{4D^3+20D}\sin x\right]$$

$$= 48\left[\frac{x}{4D(D^2+5)}\sin 3x + \frac{x}{4D(D^2+5)}\sin x\right]$$

$$= 48\left[\frac{x}{4D}\frac{\sin 3x}{(-1)(3)^2+5} + \frac{x}{4D}\frac{\sin x}{[(-1)(1)^2+5]}\right]$$

$$= 48\left[\frac{x}{4D}\frac{\sin 3x}{-4} + \frac{x}{4D}\frac{\sin x}{4}\right]$$

$$= \frac{48x}{16}\left[\frac{-1}{D}\sin 3x + \frac{1}{D}\sin x\right]$$

$$= 3x\left[\frac{-(-\cos 3x)}{3} - \cos x\right]$$

$$= x\cos 3x - 3x\cos x$$

iii) ∴ C. S. is

$$y = C_1\cos x + C_2\sin x + C_3\cos 3x + C_4\sin 3x + x\cos 3x - 3x\cos x \quad\text{... ... (1)}$$
$$y' = -C_1\sin x + C_2\cos x - 3C_3\sin 3x + 3C_4\cos 3x - 3x\sin 3x$$
$$+ \cos 3x + 3x\sin x - 3\cos x \quad\text{... ... (2)}$$
$$y'' = -C_1\cos x - C_2\sin x - 9C_3\cos 3x - 9C_4\sin 3x - 9x\cos 3x - 3\sin 3x$$
$$- 3\sin 3x + 3x\cos x + 3\sin x + 3\sin x \quad\text{... ... (3)}$$
$$y''' = C_1\sin x - C_2\cos x + 27C_3\sin 3x - 27C_4\cos 3x + 27x\sin 3x - 9\cos 3x$$
$$-9\cos 3x - 9\cos 3x + 3x\sin x + 3\cos x + 3\cos x + 3\cos x \quad\text{... ... (4)}$$

Given $x = 0, y = 0$

Equation(1) becomes, $0 = C_1 + C_3$ ... ... (a)

Given $x = 0, \ y' = 0$

Equation(2) becomes, $0 = C_2 + 3C_4 + 1 - 3$

∴ $3C_4 + C_2 = 2$ ... ... (b)

Given $x = 0, \ y'' = -8$

Equation(3) becomes, $-8 = -C_1 - 9C_3$

∴ $C_1 + 9C_3 = 8$ ... ... (c)

Given $x = 0, \ y''' = -18$

Equation(4) becomes, $-18 = -C_2 - 27C_4 - 9 - 9 - 9 + 3 + 3 + 3$

$$-18 = -C_2 - 27C_4 - 18$$
$$-18 + 18 = -C_2 - 27C_4$$

$$0 = -C_2 - 27C_4$$

$\therefore C_2 + 27\ C_4 = 0$ ... ... (d)

Now,   Equation(a) $-$ (c); $C_3 - 9C_3 = 0 - 8$

$$-8C_3 = -8$$

$C_3 = 1$

$\therefore$ Equation(a) becomes, $C_1 = -1$

Now,   Equation(b) $-$ (d) $3C_4 - 27C_4 = 2 - 0$

$$-24C_4 = 2$$

$$\therefore C_4 = -\frac{1}{12}$$

Equation (b) becomes, $3\left(\dfrac{-1}{12}\right) + C_2 = 2$; $C_2 = 2 + \dfrac{1}{4}$; $\therefore C_2 = \dfrac{9}{4}$

$\therefore$ **Required solution is , equation(1) becomes,**

$$y = -\cos x + \frac{9}{4}\sin x + \cos 3x - \frac{1}{12}\sin 3x + x\cos 3x - 3x\cos x$$

Example 34: Solve $\dfrac{d^2y}{dx^2} + y = \sin x.\sin 2x$

Solution: Given D. E. is in symbolic form as $(D^2 + 1)y = \sin x.\sin 2x$

   C. S. = C. F. + P. I.

i) To find C. F.

Its A. E. is $D^2 + 1 = 0$; $D^2 = -1$; $\therefore D = \pm i$

$\therefore$ C. F. = $c_1 \cos x + c_2 \sin x$

ii) To find P. I.

$$
\begin{aligned}
\text{P. I.} &= \frac{1}{D^2+1}\sin x.\sin 2x\left\{\because \sin A - \sin B = -\frac{1}{2}(\cos(A+B) - \cos(A-B))\right. \\
&= \frac{1}{D^2+1}\left[\frac{-1}{2}(\cos 3x - \cos x)\right] \\
&= \frac{-1}{2}\left[\frac{1}{D^2+1}\cos x - \frac{1}{D^2+1}\cos x\right] \\
&= \frac{-1}{2}\left[\frac{1.\cos 3x}{(-1)(3)^2+1} - \frac{x}{2D}\cos x\right] \\
&= \frac{1}{16}\cos 3x + \frac{x}{4}\sin x
\end{aligned}
$$

iii) $\therefore$ C. S. is $y = c_1 \cos x + c_2 \sin x + \dfrac{1}{16}\cos 3x + \dfrac{x}{4}\sin x$

6. iv   Examples on:

Type IV: $X = x^m$, m is a positive integer

$$\frac{1}{f(D)}x^m = \frac{1}{1 \pm z}x^m, \quad \text{where z is function of } D = (1 \pm z)^{-1}x^m$$

Note: 1) $(1+z)^{-1} = 1 - z + z^2 - z^3 + \cdots$

       2) $(1-z)^{-1} = 1 + z + z^2 + z^3 + \cdots$

3) $(1+z)^n = 1 + nz + \dfrac{n(n-1)}{2!}z^2 + \cdots$

4) $(1+z)^{-2} = 1 - 2z + 3z^2 - 4z^3 + \cdots$

5) $(1-z)^{-2} = 1 + 2z + 3z^2 + 4z^3 + \cdots$

Example 35: Solve $\left(D^2 + 5D + 4\right)y = x^2 + e^x$

Solution: Given D. E. is $(D^2 + 5D + 4)y = x^2 + e^x$

   C. S. = C. F. + P. I.

i) To find C. F.

Its A. E. is $D^2 + 5D + 4 = 0$; $(D+1)(D+4) = 0$; $\therefore D = -1, -4$

$\therefore$ C. F. $= C_1 e^{-x} + C_2 e^{-4x}$

ii) To find P. I.

$$\text{P. I.} = \frac{1}{D^2 + 5D + 4}(x^2 + e^x)$$

$$= \frac{1}{D^2 + 5D + 4}x^2 + \frac{1}{D^2 + 5D + 4}e^x$$

$$= \frac{1}{4\left(1 + \frac{D^2 + 5D}{4}\right)}x^2 + \frac{e^x}{1^2 + 5(1) + 4}$$

$$= \frac{1}{4}\left(1 + \frac{D^2 + 5D}{4}\right)^{-1}x^2 + \frac{e^x}{10}$$

$$= \frac{1}{4}\left[1 - \frac{D^2 + 5D}{4} + \left(\frac{D^2 + 5D}{4}\right)^2 - \cdots\right]x^2 + \frac{e^x}{10}$$

$$= \frac{1}{4}\left[x^2 - \frac{1}{4}(2 + 10x)\frac{1}{16}25(2)\right] + \frac{e^x}{10}$$

$$= \frac{1}{4}\left(x^2 - \frac{1}{2} - \frac{5x}{2} + \frac{25}{8}\right) + \frac{e^x}{10}$$

$$= \frac{1}{4}\left(x^2 - \frac{5x}{2} + \frac{21}{8}\right) + \frac{e^x}{10}$$

iii) ∴ C.S. is $y = C_1 e^{-x} + C_2 e^{-4x} + \dfrac{1}{4}\left(x^2 - \dfrac{5x}{2} + \dfrac{21}{8}\right) + \dfrac{e^x}{10}$

Example 36: Solve $\left(D^2 - 2D + 4\right) y = 3x^2 - 5x + 2$

Solution: Given D. E. is $(D^2 - 2D + 4) y = 3x^2 - 5x + 2$

  C. S. = C. F. + P. I.

i) To find C. F.

Its A. E. is $D^2 - 2D + 4 = 0$

$$D = \frac{2 \pm \sqrt{4 - 16}}{2} = 1 \pm i\sqrt{3}$$

∴ C.F. = $e^x(C_1 \cos \sqrt{3}x + C_2 \sin\sqrt{3}x)$

ii) To find P. I.

$$\begin{aligned}
\text{P.I.} &= \frac{1}{D^2 - 2D + 4}(3x^2 - 5x + 2) \\[2mm]
&= \frac{1}{4\left(1 + \frac{D^2 - 2D}{4}\right)}(3x^2 - 5x + 2) \\[2mm]
&= \frac{1}{4}\left(1 + \frac{D^2 - 2D}{4}\right)^{-1}(3x^2 - 5x + 2) \\[2mm]
&= \frac{1}{4}\left[1 - \frac{D^2 - 2D}{4} + \left(\frac{D^2 - 2D}{4}\right)^2 - \cdots\right](3x^2 - 5x + 2) \\[2mm]
&= \frac{1}{4}\left[(3x^2 - 5x + 2) - \frac{1}{4}(6 - 12x + 10) + \frac{1}{16}\,4.6\right] \\[2mm]
&= \frac{1}{4}\left[3x^2 - 5x + 2 - \frac{3}{2} + 3x - \frac{5}{2} + \frac{3}{2}\right] \\[2mm]
&= \frac{1}{4}\left[3x^2 - 2x - \frac{1}{2}\right]
\end{aligned}$$

iii) ∴ C. S. is $y = e^x(C_1 \cos \sqrt{3}\,x + C_2 \sin \sqrt{3}x) + \dfrac{1}{4}\left(3x^2 - 2x - \dfrac{1}{2}\right)$

Example 37: Solve $\dfrac{d^2y}{dx^2} + 4y = \sin 3x + e^x + x^2$

Solution: Given D. E. in symbolic form as $(D^2 + 4)y = \sin 3x + e^x + x^2$

  C. S. = C. F. + P. I.

i) To find C. F.

Its A. E. is $D^2 + 4 = 0;\ D^2 = -4$

∴ $D = \pm 2i$

∴ C. F. = $C_1 \cos 2x + C_2 \sin 2x$

ii) To find P.I.

$$P.I. = \frac{1}{D^2 + 4}(\sin 3x + e^x + x^2)$$

$$= \frac{1}{D^2 + 4}\sin 3x + \frac{1}{D^2 + 4}e^x + \frac{1}{D^2 + 4}x^2$$

$$= \frac{\sin 3x}{(-1)(3)^2 + 4} + \frac{e^x}{1^2 + 4} + \frac{1}{4\left(1 + \frac{D^2}{4}\right)}x^2$$

$$= \frac{\sin 3x}{-5} + \frac{e^x}{5} + \frac{1}{4}\left(1 + \frac{D^2}{4}\right)^{-1}x^2$$

$$= -\frac{1}{5}\sin 3x + \frac{e^x}{5} + \frac{1}{4}\left(1 - \frac{D^2}{4} + \left(\frac{D^2}{4}\right)^2 - \cdots\right)x^2$$

$$= -\frac{1}{5}\sin 3x + \frac{e^x}{5} + \frac{1}{4}\left(x^2 - \frac{2}{4} + \cdots\right)$$

$$= \frac{-\sin 3x}{5} + \frac{e^x}{5} + \frac{x^2}{4} - \frac{1}{8}$$

iii) ∴ C. S. is $y = C_1 \cos 2x + C_2 \sin 2x - \dfrac{\sin 3x}{5} + \dfrac{e^x}{5} + \dfrac{x^2}{4} - \dfrac{1}{8}$

Example 38: Solve $(D - 2)^2 y = 8\left(e^{2x} + \sin 2x + x^2\right)$

Solution: Given D. E. is $(D - 2)^2 y = 8\left(e^{2x} + \sin 2x + x^2\right)$

C. S. = C. F. + P. I.

i) To find C. F.

Its A. E. is $(D - 2)^2 = 0$;   ∴ $D = 2, \ 2$

∴ C. F. = $(C_1 + C_2 x)e^{2x}$

ii) To find P. I.

$$P.I. = \frac{1}{(D - 2)^2}\, 8(e^{2x} + \sin 2x + x^2)$$

$$= 8\left[\frac{1}{(D - 2)^2}e^{2x} + \frac{1}{(D - 2)}\sin 2x + \frac{1}{(D - 2)^2}x^2\right]$$

$$= 8\left[\frac{x^2}{2!}e^{2x} + \frac{1}{D^2 - 4D + 4}\sin 2x + \frac{1}{D^2 - 4D + 4}x^2\right]$$

$$= 8\left[\frac{x^2}{2}e^{2x} + \frac{1}{(-1)(2)^2 - 4D + 4}\sin 2x + \frac{1}{4\left(1 + \frac{D^2 - 4D}{4}\right)}x^2\right]$$

$$= 8\left[\frac{x^2}{2}e^{2x} + \frac{1}{-4D}\sin 2x + \frac{1}{4}\left(1 + \frac{D^2 - 4D}{4}\right)^{-1}x^2\right]$$

$$= 8\left[\frac{x^2}{2}e^{2x} - \frac{1}{4}\int \sin 2x \ dx + \frac{1}{4}\left(1 - \left(\frac{D^2 - 4D}{4}\right) + \frac{16D^2}{16}\right)x^2\right]$$

$$= 8\left[\frac{x^2}{2}e^{2x} - \frac{1}{4}\left(\frac{-\cos 2x}{2}\right) + \frac{1}{4}\left(x^2 - \frac{1}{4}(2 - 4(2x)) + 2\right)\right]$$

$$= 8\left[\frac{x^2 - e^{2x}}{2} + \frac{1}{8}\cos 2x + \frac{1}{4}\left(x^2 - \frac{1}{2} + 2x + 2\right)\right]$$

$$= 8\left[\frac{x^2 e^{2x}}{2} + \frac{\cos 2x}{8} + \frac{x^2}{4} + \frac{x}{2} + \frac{3}{8}\right]$$

$$= 4x^2 e^{2x} + \cos 2x + 2x^2 + 4x + 3$$

iii) ∴ C. S. is $y = (C_1 + C_2 x)e^{2x} + 4x^2 e^{2x} + \cos 2x + 2x^2 + 4x + 3$

Example 39: Solve $(D^2 - 1)y = e^x + x^2$, **with** $y = \dfrac{dy}{dx} = 0$ **at** $x = 0$

Solution: Given D. E. is $(D^2 - 1)y = e^x + x^2$

  C. S. = C. F. + P. I.

i) To find C. F.

Its A. E. is $(D^2 - 1) = 0$; $D^2 = 1$; ∴ $D = \pm 1$

∴ C. F. $= C_1 e^x + C_2 e^{-x}$

ii) To find P. I.

$$\text{P. I.} = \frac{1}{D^2 - 1}(e^x + x^2)$$

$$= \frac{1}{D^2 - 1}e^x + \frac{1}{D^2 - 1}x^2$$

$$= \frac{x}{2D}e^x + \frac{1}{-(1 - D^2)}x^2$$

$$= \frac{x}{2}\int e^x \ dx - (1 - D^2)^{-1}x^2$$

$$= \frac{x}{2}e^x - (1 + D^2 + (D^2)^2 \ldots)x^2$$

$$= \frac{x}{2}e^x - (x^2 + 2)$$

iii) ∴ C. S. is $y = C_1 e^x + C_2 e^{-x} + \dfrac{x}{2}e^x - x^2 - 2$ …… (1)

Given x = 0, y = 0

Equation (1) → $0 = C_1 + C_2 - 2$

∴ $C_1 + C_2 = 2$ …… (a)

Now,   Different equation(1) w. r. t. 'x'

∴ $y' = C_1 e^x - C_2 e^{-x} + \dfrac{x}{2}e^x + \dfrac{e^x}{2} - 2x$ …… (2)

Given $x = 0, \ y' = 0$

Equation (2) $\rightarrow$ $0 = C_1 - C_2 + \dfrac{1}{2}$

$\therefore C_1 - C_2 = \dfrac{-1}{2}$ ... ... (b)

Now, Adding equation(a) and (b), $C_1 + C_1 = 2 - \dfrac{1}{2}$; $2C_1 = \dfrac{3}{2}$; $\boldsymbol{C_1 = \dfrac{3}{4}}$

Put $C_1 = \dfrac{3}{4}$ in equation(a), $C_2 = 2 - \dfrac{3}{4} = \dfrac{8-3}{4} = \dfrac{5}{4}$; $\boldsymbol{C_2 = \dfrac{5}{4}}$

$\therefore$ **Required particular solution is**

Equation (1) becomes, $y = \dfrac{3}{4}e^x + \dfrac{5}{4}e^{-x} + \dfrac{3}{2}e^x - x^2 - 2$

Example 40: **Solve $\left(D^4 - 2D^3 + D^2\right)y = x^3$**

Solution: Given D. E. is $(D^4 - 2D^3 + D^2)y = x^3$

   C. S. = C. F. + P. I.

i) To find C. F.

Its A. E. is $D^4 - 2D^3 + D^2 = 0$

$D^2(D^2 - 2D + 1) = 0$

$D^2(D - 1)^2 = 0$;

$D^2 = 0$ or $(D - 1)^2 = 0$; $\therefore D = 0, 0, 1, 1$

$\therefore$ C. F. = $C_1 + C_2 x + (C_3 + C_4 x)e^x$

ii) To find P. I.

P. I. $= \dfrac{1}{D^4 - 2D^3 + D^2} x^3$

$= \dfrac{1}{D^2(D^2 - 2D + 1)} x^3$

$= \dfrac{1}{D^2}[1 + (D^2 - 2D)]^{-1} x^3$

$= \dfrac{1}{D^2}[1 - (D^2 - 2D) + (D^2 - 2D)^2 - (D^2 - 2D)^3 + \cdots] x^3$

$= \dfrac{1}{D^2}[1 - D^2 + 2D - 4D^3 + 4D^2 + 8D^3 + \cdots] x^3$

$= \dfrac{1}{D^2}[1 + 2D + 3D^2 + 4D^3 + \cdots] x^3$

$= \dfrac{1}{D^2}[x^3 + 2.3x^2 + 3.6x + 4.6]$

$= \dfrac{1}{D^2}[x^3 + 6x^2 + 18x + 24]$ $= \dfrac{x^5}{20} + 6\dfrac{x^4}{12} + 18\dfrac{x^3}{6} + 24\dfrac{x^2}{2}$

$$= \frac{x^5}{20} + \frac{x^4}{2} + 3x^3 + 12\, x^2$$

iii) ∴ C. S. is $y = C_1 + C_2x + (C_3 + C_4x)e^x + \dfrac{x^5}{20} + \dfrac{x^4}{2} + 3x^3 + 12x^2$

---

6. v   Examples on:

Type V: $X = e^{ax}.V,$ where V is a function of x

$$\frac{1}{f(D)}e^{ax}.V = e^{ax}\frac{1}{f(D+a)}V$$

Example 41: Solve $(D^3 + 3D^2 - 4)y = xe^{-2x}$

Solution: Given D. E. is $(D^3 + 3D^2 - 4)y = xe^{-2x}$

   C. S. = C. F. + P. I.

i) To find C. F.

Its A. E. is $D^3 + 3D^2 - 4 = 0$;   ∴ $D = 1, -2, -2$

∴ C. F. $= C_1e^x + (C_2 + C_3x)e^{-2x}$

ii) To find P. I.

$$\text{P. I.} = \frac{1}{D^3 + 3D^2 - 4}xe^{-2x}$$

$$= \frac{e^{-2x}}{(D-2)^3 + 3(D-2)^2 - 4}.x$$

$$= \frac{e^{-2x}}{D^3 - 3D^2(2) + 3D(2)^2 - 2^3 + 3(D^2 - 4D + 4) - 4}x$$

$$= \frac{e^{-2x}}{D^3 - 6D^2 + 12D - 3 + 3D^2 - 12D + 12 - 4}x$$

$$= \frac{e^{-2x}}{D^3 - 3D^2}x$$

$$= \frac{e^{-2x}}{-3D^2\left(1 - \frac{D^3}{3D^2}\right)}x$$

$$= \frac{-e^{-2x}}{3}\frac{1}{D^2}\left(1 - \frac{D}{3}\right)^{-1}x$$

$$= \frac{-e^{-2x}}{3}\frac{1}{D^2}\left(1 + \frac{D}{3} + \cdots\right)x$$

$$= \frac{-e^{-2x}}{3}\frac{1}{D^2}\left(x + \frac{1}{3}\right)$$

$$= \frac{-e^{-2x}}{3}\left(\frac{x^3}{6} + \frac{1}{3}\frac{x^2}{2}\right) \qquad = \frac{-e^{-2x}}{3}\left(\frac{x^3}{6} + \frac{x^2}{6}\right)$$

$$= -\frac{1}{18}e^{-2x}(x^3 + x^2)$$

iii) $\therefore$ C.S. is $y = C_1 e^x + (C_2 + C_3 x)e^{-2x} - \frac{1}{18}e^{-2x}(x^3 + x^2)$

Example 42: Solve $(D^2 - 2D + 2)\,y = e^x \tan x + 3x$

Solution: Given D. E. is $(D^2 - 2D + 2)\,y = e^x \tan x + 3x$

C. S. = C. F. + P. I.

i) To find C. F.

Its A. E. is $D^2 - 2D + 2 = 0$;

$$D = \frac{2 \pm \sqrt{4 - 8}}{2} = \frac{2 \pm 2i}{2}; \qquad \therefore D = 1 \pm i$$

$\therefore$ C. F. $= e^x(C_1 \cos x + C_2 \sin x)$

ii) To find P. I.

P. I. $= \dfrac{1}{D^2 - 2D + 2}(e^x \tan x + 3x)$

$= \dfrac{1}{D^2 - 2D + 2} e^x \cdot \tan x + 3\dfrac{1}{D^2 - 2D + 2} x$

$= \dfrac{e^x}{(D+1)^2 - 2(D+1) + 2} \tan x + 3\dfrac{1}{2\left(1 + \frac{D^2 - 2D}{2}\right)} x$

$= \dfrac{e^x}{D^2 + 2D + 1 - 2D - 2 + 2} \tan x + \dfrac{3}{2}\left(1 + \dfrac{D^2 - 2D}{2}\right)^{-1} x$

$= \dfrac{e^x}{D^2 + 1} \tan x + \dfrac{3}{2}\left(1 - \dfrac{D^2 - 2D}{2} + \cdots \right) x$

$= \dfrac{e^x}{(D+i)(D-i)} \cdot \tan x + \dfrac{3}{2}\left(x - \dfrac{1}{2}(-2)\right)$

$= e^x \dfrac{1}{2i}\left[\dfrac{1}{D-i} - \dfrac{1}{D+i}\right] \tan x + \dfrac{3}{2}(x + 1)$

$= \dfrac{e^x}{2i}\left[\dfrac{1}{D-i} \tan x - \dfrac{1}{D+i} \tan x\right] + \dfrac{3}{2}(x + 1)$

$= \dfrac{e^x}{2i}\left[e^{ix}\int e^{-ix} \tan x\, dx - e^{-ix}\int e^{ix} \tan x\, dx\right] + \dfrac{3}{2}(x + 1)$

$= \dfrac{e^x}{2i}\left[e^{ix}\int (\cos x - i\sin x) \tan x\, dx - e^{-ix}\int (\cos x + i\sin x) \tan x\, dx\right] + \dfrac{3}{2}(x + 1)$

$\{\because$ Eulers formula $e^{i\theta} = \cos\theta + i\sin\theta\}$

$= \dfrac{e^x}{2i}\left[e^{ix}\int \left(\sin x - \dfrac{i\,\sin^2 x}{\cos x}\right) dx - e^{-ix}\int \left(\sin x + i\dfrac{\sin^2 x}{\cos x}\right) dx\right] + \dfrac{3}{2}(x + 1)$

$$= \frac{e^x}{2i}\left[\left(e^{ix}-e^{-ix}\right)\int \sin x \, dx \; -i\left(e^{ix}+e^{-ix}\right)\int \left(\frac{1}{\cos x}-\cos x\right)dx\right]+\frac{3}{2}(x+1)$$

$$= \frac{e^x}{2i}\left[2i\sin x\,(-\cos x) - i\,2\cos x\,(\log|\sec x + \tan x| - \sin x)\right]+\frac{3}{2}(x+1)$$

$$= \frac{e^x}{2i}\left[-2i\sin x \cdot \cos x - 2i\cos x \cdot \log|\sec x + \tan x| + 2i\sin x \cos x\right]+\frac{3}{2}(x+1)$$

$$= \frac{e^x}{2i}\left(-2i\cos x \cdot \log|\sec x + \tan x|\right)+\frac{3}{2}(x+1)$$

$$= -e^x \cdot \cos x \cdot \log|\sec x + \tan x|+\frac{3}{2}(x+1)$$

iii) $\therefore$ C. S. is $y = e^x(C_1\cos x + C_2\sin x) - e^x\cos x \cdot \log|\sec x + \tan x| + \frac{3}{2}(x+1)$

Example 43: Solve $\left(D^2 + 4\right)y = e^x\sin^2\dfrac{x}{2}$

Solution: The given D. E. is $(D^2 + 4)\,y = e^x\sin^2\dfrac{x}{2}$

  C. S. = C. F. + P. I.

i) To find C. F.

Its A. E. is $D^2 + 4 = 0$; $D^2 = -4$; $\therefore D = \pm\,2i$

$\therefore$ C. F. $= C_1\cos 2x + C_2\sin 2x$

ii) To find P. I.

$$\text{P. I.} \; = \frac{1}{D^2 + 4}e^x \cdot \sin^2\frac{x}{2}$$

$$= \frac{e^x}{(D + 1)^2 + 4}\sin^2\frac{x}{2}$$

$$= \frac{e^x}{D^2 + 2D + 1 + 4}\left(\frac{1 - \cos x}{2}\right)\left\{\because \sin^2\theta = \frac{1 - \cos\theta}{2}\right.$$

$$= \frac{e^x}{2}\left[\frac{1}{D^2 + 2D + 5}e^{0x} - \frac{1}{D^2 + 2D + 5}\cos x\right]$$

$$= \frac{e^x}{2}\left[\frac{1}{0 + 0 + 5} - \frac{1}{(-1)(1)^2 + 2D + 5}\cos x\right]$$

$$= \frac{e^x}{2}\left[\frac{1}{5} - \frac{1}{2D + 4}\cos x\right]$$

$$= \frac{e^x}{2}\left[\frac{1}{5} - \frac{2D - 4}{4D^2 - 16}\cos x\right]$$

$$= \frac{e^x}{2}\left[\frac{1}{5} - (2D - 4)\frac{\cos x}{4(-1)(1)^2 - 16}\right]$$

$$= \frac{e^x}{2}\left[\frac{1}{5} + \frac{1}{20}(2D\cos x - 4\cos x)\right]$$

$$= \frac{e^x}{2}\left[\frac{1}{5} + \frac{1}{20}2(-\sin x) - \frac{4\cos x}{20}\right]$$

$$= \frac{e^x}{2}\left[\frac{1}{5} - \frac{1}{10}\sin x - \frac{1}{5}\cos x\right]$$

iii) ∴ C. S. is $y = C_1 \cos 2x + C_2 \sin 2x + \dfrac{e^x}{2}\left[\dfrac{1}{5} - \dfrac{1}{10}\sin x - \dfrac{1}{5}\cos x\right]$

Example 44: Solve $(D^3 - D^2 + 3D + 5)y = e^x.\cos 3x$

Solution: The given D. E. is $(D^3 - D^2 + 3D + 5)y = e^x.\cos 3x$

   C. S. = C. F. + P. I.

i) To find C. F.

   Its A. E. is $D^3 - D^2 + 3D + 5 = 0$

By Synthetic division

$$
\begin{array}{r|rrrr}
-1 & 1 & -1 & 3 & 5 \\
   &   & -1 & 2 & -5 \\
\hline
   & 1 & -2 & 5 & 0
\end{array}
$$

$(D + 1)(D^2 - 2D + 5) = 0$

$$D = -1, \quad \frac{2 \pm \sqrt{4 - 20}}{2}; \quad D = -1, \ 1 \pm 2i$$

∴ C. F. $= C_1 e^{-x} + e^x(C_2 \cos 2x + C_3 \sin 2x)$

ii) To find P. I.

$$\text{P. I.} = \frac{1}{D^3 - D^2 + 3D + 5}e^x.\cos 3x$$

$$= e^x \frac{1}{(D+1)^3 - (D+1)^2 + 3(D+1) + 5}\cos 3x$$

$$= e^x \frac{1}{D^2 + 3D^2(1) + 3D(1)^2 + 1 - (D^2 + 2D + 1) + 3D + 3 + 5}\cos 3x$$

$$= e^x \frac{1}{D^3 + 3D^2 + 3D + 1 - D^2 - 2D - 1 + 3D + 8}\cos 3x$$

$$= e^x \frac{1}{D^3 + 2D^2 + 4D + 8}\cos 3x$$

$$= e^x \frac{1}{(-1)(3)^2 D + 2(-1)(3)^2 + 4D + 8}\cos 3x$$

$$= e^x \frac{1}{-9D - 18 + 4D + 8}\cos 3x$$

$$= e^x \frac{1}{-10 - 5D}\cos 3x$$

$$= e^x \frac{1}{-(10 + 5D)}\cos 3x$$

$$= -e^x \frac{10 - 5D}{100 - 25D^2} \cos 3x$$

$$= -e^x \frac{(10 - 5D)}{100 - 25(-1)(3)^2} \cdot \cos 3x$$

$$= \frac{-e^x}{325}(10 \cos 3x - 5D \cos 3x)$$

$$= \frac{-e^x}{325}(10 \cos 3x + 15 \sin 3x)$$

$$= \frac{-e^x}{65}(2 \cos 3x + 3 \sin 3x)$$

iii) ∴ C.S. is $y = C_1 e^{-x} + e^x(C_2 \cos 2x + C_3 \sin 2x) - \frac{e^x}{65}(2 \cos 3x + 3 \sin 3x)$

Example 45: Solve $\left(\frac{d^2}{dx^2} + n^2 \right)y = x^2 e^x$

Solution: Given D. E. in symbolic form as $(D^2 + n^2)y = x^2 e^x$

  C. S. = C. F. + P. I.

i) To find C. F.

Its A. E. is $D^2 + n^2 = 0$; $D^2 = -n^2$;   ∴ $D = \pm$ ni

∴ C. F. = $C_1 \cos nx + C_2 \sin nx$

ii) To find P. I.

$$P.I. = \frac{1}{D^2 + n^2}x^2 e^x$$

$$= e^x \frac{1}{(D + 1)^2 + n^2}x^2$$

$$= e^x \frac{1}{D^2 + 2D + 1 + n^2}x^2$$

$$= e^x \frac{1}{(1 + n^2)\left(1 + \frac{D^2 + 2D}{1 + n^2}\right)}x^2$$

$$= \frac{e^x}{1 + n^2}\left(1 + \frac{D^2 + 2D}{1 + n^2}\right)^{-1}x^2$$

$$= \frac{e^x}{1 + n^2}\left(1 - \frac{D^2 + 2D}{1 + n^2} + \frac{4D^2}{(1 + n^2)^2} - \cdots\right)x^2$$

$$= \frac{e^x}{1 + n^2}\left(x^2 - \frac{2 + 4x}{n^2 + 1} + \frac{8}{(n^2 + 1)^2}\right)$$

iii) ∴ C. S. is $y = C_1 \cos nx + C_2 \sin nx + \frac{e^x}{1 + n^2}\left(x^2 - \frac{2 + 4x}{n^2 + 1} + \frac{8}{(n^2 + 1)^2}\right)$

Example 46: Solve $\dfrac{d^3y}{dx^3} - 7\dfrac{dy}{dx} - 6y = e^{2x}(1+x)$

Solution: Given D. E. in symbolic form as $(D^3 - 7D - 6)y = e^{2x}(1+x)$

C. S. = C. F. + P. I.

i) To find C. F.

Its A. E. is $D^3 - 7D - 6 = 0$; $\therefore D = -1, -2, 3$

$\therefore$ C. F. $= C_1e^{-x} + C_2e^{-2x} + C_3e^{3x}$

ii) To find P. I.

$$P.I. = \frac{1}{D^2 - 7D - 6}e^{2x}(1+x)$$

$$= e^{2x}\frac{1}{(D+2)^3 - 7(D+2) - 6}(1+x)$$

$$= e^{2x}\frac{1}{D^3+3D^2(2)+3D(2)^2 + 2^3 - 7D - 14 - 6}(1+x)$$

$$= e^{2x}\frac{1}{D^3+6D^2 + 5D - 12}(1+x)$$

$$= e^{2x}\frac{1}{-12\left(1 - \frac{(D^3+6D^2+5D)}{12}\right)}(1+x)$$

$$= \frac{-e^{2x}}{12}\left[1 - \frac{D^3 + 6D^2 + 5D}{12}\right]^{-1}(1+x)$$

$$= \frac{-e^{2x}}{12}\left[1 + \frac{D^3 + 6D^2 + 5D}{12} + \cdots\right](1+x)$$

$$= \frac{-e^{2x}}{12}\left[(1+x) + \frac{1}{12}5D(1+x)\right]$$

$$= \frac{-e^{2x}}{12}\left[1 + x + \frac{5}{12}(1)\right] \quad = \frac{-e^{2x}}{12}\left(x + \frac{17}{12}\right)$$

iii) $\therefore$ **C. S. is** $y = C_1e^{-x} + C_2e^{-2x} + C_3e^{3x} - \dfrac{e^{2x}}{12}\left(x + \dfrac{17}{12}\right)$

Example 47: Solve $\left(D^2 + 5D + 6\right)y = e^{-2x}.\sin 2x + 4x^2e^x$

Solution: The given D. E. is $(D^2 + 5D + 6) y = e^{-2x}\sin 2x + 4x^2e^x$

C. S. = C. F. + P. I.

i) To find C. F.

Its A. E. is $(D^2 + 5D + 6) = 0$; $(D + 2)(D + 3) = 0$; $\therefore D = -2, -3$

$\therefore$ C. F. $= C_1e^{-2x} + C_2e^{-3x}$

ii) To find P. I.

$$P.I. = \frac{1}{D^2 + 5D + 6}(e^{-3x}.\sin 2x + 4x^2 e^x)$$

$$= \frac{1}{D^2 + 5D + 6}e^{-2x}\sin 2x + \frac{1}{D^2 + 5D + 6}4x^2 e^x$$

$$= e^{-2x}\frac{1}{(D-2)^2 + 5(D-2) + 6}\sin 2x + 4e^x\frac{1}{(D+1)^2 + 5(D+1) + 6}x^2$$

$$= e^{-2x}\frac{1}{D^2 - 4D + 4 + 5D - 10 + 6}\sin 2x + 4e^x\frac{1}{D^2 + 2D + 1 + 5D + 5 + 6}x^2$$

$$= e^{-2x}\frac{1}{D^2 + D}\sin 2x + 4e^x\frac{1}{D^2 + 7D + 12}x^2$$

$$= e^{-2x}\frac{1}{(-1)(2)^2 + D}\sin 2x + 4e^x\frac{1}{12\left(1 + \frac{D^2 + 7D}{12}\right)}x^2$$

$$= e^{-2x}\frac{1}{D-4}\sin 2x + \frac{e^x}{3}\left(1 + \frac{D^2 + 7D}{12}\right)^{-1}x^2$$

$$= e^{-2x}\frac{(D+4)}{D^2 - 16}\sin 2x + \frac{e^x}{3}\left[1 - \frac{D^2 + 7D}{12} + \left(\frac{D^2 + 7D}{12}\right)^2 - \cdots\right]x^2$$

$$= e^{-2x}\frac{(D+4)\sin 2x}{-(2)^2 - 16} + \frac{e^x}{3}\left[x^2 - \frac{1}{12}(2 + 7(2x)) + \frac{1}{144}(49(2))\right]$$

$$= e^{-2x}\frac{(2\cos 2x + 4\sin 2x)}{-20} + \frac{e^x}{3}\left[x^2 - \frac{1}{6}(1 + 7x) + \frac{(49)}{72}\right]$$

$$= -\frac{e^{-2x}}{10}(\cos 2x + 2\sin 2x) + \frac{e^x}{3}\left(x^2 - \frac{1}{6} - \frac{7x}{6} + \frac{49}{72}\right)$$

$$= -\frac{e^{-2x}}{10}(\cos 2x + 2\sin 2x) + \frac{e^x}{3}\left(x^2 - \frac{7x}{6} + \frac{37}{72}\right)$$

iii) $\therefore$ C.S. is $y = C_1 e^{-2x} + C_2 e^{-3x} - \frac{e^{-2x}}{10}(\cos 2x + 2\sin 2x) + \frac{e^x}{3}\left(x^2 - \frac{7x}{6} + \frac{37}{72}\right)$

Example 48: Solve $\dfrac{d^2 y}{dx^2} - 2\dfrac{dy}{dx} + y = x\, e^x \sin x$

Solution: The given D. E. is in symbolic form is

$(D^2 - 2D + 1)y = x\, e^x \sin x$

   C.S. = C.F. + P.I.

i) **To find C. F.**

Its A. E. is $(D^2 - 2D + 1) = 0$; $(D - 1)^2 = 0$; $\therefore D = 1, 1$

$\therefore$ C.F. = $(C_1 + C_2 x)e^x$

ii) **To find P. I.**

$$\text{P.I.} = \frac{1}{D^2 - 2D + 1} e^x \cdot x \cdot \sin x$$

$$= \frac{1}{(D-1)^2} e^x \cdot x \cdot \sin x$$

$$= e^x \frac{1}{(D+1-1)^2} x \cdot \sin x$$

$$= e^x \frac{1}{D^2} x \cdot \sin x \quad = e^x \frac{1}{D}\left[\frac{1}{D} x \cdot \sin x\right] \quad = e^x \frac{1}{D} \int x \sin x \, dx$$

$$= e^x \frac{1}{D}[x(-\cos x) - 1.(-\sin x)] \qquad \{\because \text{By LIATE rule}$$

$$= e^x \frac{1}{D}[-x \cos x + \sin x]$$

$$= e^x[-x.\sin x + 1.(-\cos x) + (-\cos x)]$$

$$= -e^x[x \sin x + \cos x + \cos x]$$

$$= -e^x[x \sin x + 2 \cos x]$$

iii) ∴ C.S. is $y = (C_1 + C_2 x)e^x - e^x(x \sin x + 2 \cos x)$

Example 49: Solve $(D^2 - 6D + 9) y = \dfrac{e^{3x}}{x^2}$

Solution: Given D.E. is $(D^2 - 6D + 9) y = \dfrac{e^{3x}}{x^2}$

C.S. = C.F. + P.I.

i) **To find C.F.**

Its A.E. is $D^2 - 6D + 9 = 0$; $(D-3)^2 = 0$;   ∴ $D = 3,3.$

∴ C.F. = $(C_1 + C_2 x)e^x$

ii) **To find P.I.**

$$\text{P.I.} = \frac{1}{D^2 - 6D + 5} \frac{e^{3x}}{x^2}$$

$$= \frac{1}{(D-3)^2} \frac{e^{3x}}{x^2}$$

$$= e^{3x} \frac{1}{(D+3-3)^2} \frac{1}{x^2}$$

$$= e^{3x} \frac{1}{D^2} \frac{1}{x^2} \quad = e^{3x} \frac{1}{D} \int \frac{1}{x^2} \, dx = e^{3x} \frac{1}{D}\left(\frac{-1}{x}\right) \quad = e^{3x} \int \frac{1}{x} \, dx$$

$$= -e^{3x} \log x$$

iii) ∴ C.S. is $y = (C_1 + C_2 x)e^{3x} - e^{3x} \log x$

Example 50: Solve $(D^2 - 4D + 4)y = 8x^2 e^{2x} \sin 2x$

Solution: Given D. E. is $(D^2 - 4D + 4) y = 8x^2 e^{2x} \sin 2x$

C. S. = C. F. + P. I.

i) To find C. F.

Its A. E. is $D^2 - 4D + 4 = 0$; $(D - 2)^2 = 0$; $\therefore D = 2, 2$

$\therefore$ C. F. $= (C_1 + C_2 x) e^{2x}$

ii) To find P. I.

$$P. I. = \frac{1}{D^2 - 4D + 4} 8x^2 e^{2x}. \sin 2x$$

$$= \frac{1}{(D - 2)^2} 8x^2 e^{2x} \sin 2x$$

$$= 8 e^{2x} \frac{1}{(D + 2 - 2)^2} x^2. \sin 2x \quad = 8 e^{2x} \frac{1}{D^2} x^2 \sin 2x$$

$$= 8 e^{2x} \frac{1}{D} \int x^2. \sin 2x \, dx$$

$$= 8 e^{2x} \frac{1}{D} \left[x^2 \left(-\frac{\cos 2x}{2} \right) - 2x \left(\frac{-\sin 2x}{4} \right) + 2 \left(\frac{\cos 2x}{8} \right) \right]$$

$$= 8 e^{2x} \frac{1}{D} \left[\frac{-1}{2} x^2 \cos 2x + \frac{1}{2} x \sin 2x + \frac{1}{4} \cos 2x \right]$$

$$= 8 e^{2x} \left[\frac{-1}{2} \int x^2 \cos 2x \, dx + \frac{1}{2} \int x \sin 2x \, dx + \frac{1}{4} \int \cos 2x \, dx \right]$$

$$= 8 e^{2x} \left\{ \frac{-1}{2} \left[x^2 \left(\frac{\sin 2x}{2} \right) - 2x \left(\frac{-\cos 2x}{4} \right) + 2 \left(\frac{-\sin 2x}{8} \right) \right] \right.$$

$$\left. + \frac{1}{2} \left[x \left(-\frac{\cos 2x}{2} \right) - \left(\frac{-\sin 2x}{4} \right) \right] + \frac{1}{4} \frac{\sin 2x}{2} \right\}$$

$$= 8 e^{2x} \left[\frac{-x^2}{4} \sin 2x - \frac{x}{4} \cos 2x + \frac{1}{8} \sin 2x - \frac{x}{4} \cos 2x + \frac{\sin 2x}{8} + \frac{\sin 2x}{8} \right]$$

$$= 8 e^{2x} \left[\left(\frac{-x^2}{4} + \frac{1}{8} + \frac{1}{8} + \frac{1}{8} \right) \sin 2x + \left(\frac{-x}{4} - \frac{x}{4} \right) \cos 2x \right]$$

$$= 8 e^{2x} \left[\left(\frac{-2x^2 + 3}{8} \right) \sin 2x + \left(\frac{-2x}{4} \right) \cos 2x \right]$$

$$= e^{2x} [(-2x^2 + 3) \sin 2x - 4x. \cos 2x]$$

$$= e^{2x} [(3 - 2x^2) \sin 2x - 4x \cos 2x]$$

iii) $\therefore$ **C. S. is** $y = (C_1 + C_2 x) e^{2x} + e^{2x} [(3 - 2x^2) \sin 2x - 4x \cos 2x]$

6. vi Examples on:

TypeVI: X = x . V,   where V is a function of x

$$\frac{1}{f(D)} X.V = X\frac{1}{f(D)}V - \frac{f'(D)}{[f(D)]^2}V$$

Example 51: **Solve** $(D^2 + 3D + 2)\, y = x \sin 2x$

Solution: Given D. E. is $(D^2 + 3D + 2)\, y = x \sin 2x$

  C. S. = C. F. + P. I.

i) To find C. F.

Its A. E. is $D^2 + 3D + 2 = 0;$ $(D + 1)(D + 2) = 0;$ $\therefore D = -1, -2$

$\therefore$ C. F. = $C_1 e^{-x} + C_2 e^{-2x}$

ii) To find P. I.

$$P.I. = \frac{1}{D^2 + 3D + 2} \, x.\sin 2x \left\{ \because \frac{1}{f(D)} X.V = X\frac{1}{f(D)}V - \frac{f'(D)}{[f(D)]^2}V \right.$$

$$= x \frac{1}{D^2 + 3D + 2} \sin 2x - \frac{(2D + 3)}{(D^2+3D + 2)^2} \sin 2x$$

$$= x. \frac{1}{(-1)(2)^2 + 3D + 2} \sin 2x - \frac{(2D + 3)}{[(-1)(2)^2 + 3D + 2]^2} \sin 2x$$

$$= x \frac{1}{3D - 2} \sin 2x - \frac{(2D + 3)}{(3D - 2)^2} \sin 2x$$

$$= x \frac{3D + 2}{9D^2 - 4} \sin 2x - \frac{(2D + 3)}{9D^2 - 12D + 4} \sin 2x$$

$$= x \frac{(3D + 2)\sin 2x}{9(-1)(2)^2 - 4} - \frac{(2D + 3).1}{9(-1)(2)^2 - 12D + 4} \sin 2x$$

$$= x \frac{(3D + 2).\sin 2x}{-40} - \frac{(2D + 3)}{-32 - 12D} \sin 2x$$

$$= \frac{-x}{40} [3D(\sin 2x) + 2 \sin 2x] + \frac{1}{4}\frac{(2D + 3)}{(8 + 3D)} \sin 2x$$

$$= \frac{-x}{40} [3 \cos 2x(2) + 2 \sin 2x] + \frac{1}{4}\frac{(2D + 3)(8 - 3D)}{64 - 9D^2} \sin 2x$$

$$= \frac{-x}{40} [6 \cos 2x + 2 \sin 2x] + \frac{1}{4}\frac{(2D + 3)(8 - 3D)\sin 2x}{64 - 9(-1)(2)^2}$$

$$= \frac{-x}{40} [6 \cos 2x + 2 \sin 2x] + \frac{1}{4}\frac{(16D - 6D^2 + 24 - 9D).\sin 2x}{100}$$

$$= \frac{-x}{40} [6 \cos 2x + 2 \sin 2x] + \frac{1}{400} (16D \sin 2x - 6D^2 \sin 2x + 24 \sin 2x - 9D \sin 2x)$$

$$= -\frac{x}{20} (3 \cos 2x + \sin 2x) + \frac{1}{400} (24 \sin 2x + 7D \sin 2x - 6D^2 \sin 2x)$$

$$= -\frac{x}{20}(3\cos 2x + \sin 2x) + \frac{1}{400}(24\sin 2x + 14\cos 2x + 24\sin 2x)$$

$$= \left(\frac{-3x}{20} + \frac{14}{400}\right)\cos 2x + \left(\frac{-x}{20} + \frac{24}{400} + \frac{24}{400}\right)\sin 2x$$

$$= \left(\frac{7 - 30x}{200}\right)\cos 2x + \left(\frac{12 - 5x}{100}\right)\sin 2x$$

iii) $\therefore$ C. S. is $\quad y = C_1 e^{-x} + C_2 e^{-2x} + \left(\frac{7 - 30x}{200}\right)\cos 2x + \left(\frac{12 - 5x}{100}\right).\sin 2x$

6. vii Examples on:

Special cases:

1) $X = a^x \dfrac{1}{f(D)} a^x = \dfrac{1}{f(D)} e^{\log a^x} = \dfrac{1}{f(D)} e^{x\log a} = \dfrac{e^{x\log a}}{f(\log a)} = \dfrac{a^x}{f(\log a)}$

2) $X = x^m.\cos ax$ or $x^m \sin ax$

i) $\dfrac{1}{f(D)} x^m.\cos ax = $ R. P. of $\dfrac{1}{f(D)} e^{iax}.x^m = $ R. P. of $e^{iax}\dfrac{1}{f(D + ia)}x^m$

ii) $\dfrac{1}{f(D)} x^m.\sin ax = $ I. P. of $\dfrac{1}{f(D)} e^{iax}.x^m = $ I. P. of $e^{iax}\dfrac{1}{f(D + ia)}x^m$

3) $X = k,$ where k is constant

$\dfrac{1}{f(D)} k = k\dfrac{1}{f(D)}.1 = k\dfrac{1}{f(D)}e^{0x} = k\dfrac{e^{0x}}{f(0)},\qquad$ provided $f(0) \neq 0$

4) $\dfrac{1}{D}X = \int X\,dx$

Example 52: Solve $(D^2 + 1)\ y = a^x.e^x$

Solution: Given D. E. is $(D^2 + 1)\ y = a^x.e^x$

$\quad$ C. S. = C. F. + P. I.

i) To find C. F.

Its A. E. is $D^2 + 1 = 0$; $D^2 = -1,\ D = \pm i$

$\therefore$ C. F. = $C_1 \cos x + C_2 \sin x$

ii) To find P. I.

$\therefore\quad$ P. I. = $\dfrac{1}{D^2 + 1}a^x e^x$

$\qquad = e^x \dfrac{1}{(D + 1)^2 + 1}a^x \quad = e^x \dfrac{1}{D^2 + 2D + 2}a^x$

$$= e^x \frac{1}{(D^2 + 2D + 2)} e^{\log a \cdot x} \quad = e^x \frac{1}{(D^2 + 2D + 2)} e^{x(\log a)}$$

$$= e^x \frac{e^{(\log a)x}}{(\log a)^2 + 2(\log a) + 2} = \frac{e^x \cdot a^x}{(\log a)^2 + 2\log a + 2}$$

iii) ∴ C. S. is $y = C_1 \cos x + C_2 \sin x + \dfrac{e^x a^x}{(\log a)^2 + 2\log a + 2}$

Example 53: Solve $(D^2 + 2D + 1)y = x \cos x$

Solution: Given D. E. is $(D^2 + 2D + 1)y = x \cos x$

  C. S. = C. F. + P. I.

i) To find C. F.

Its A. E. is $D^2 + 2D + 1 = 0$; $(D + 1)^2 = 0$;   ∴ $D = -1, -1$

∴ C. F. = $(C_1 + C_2 x)e^{-x}$

ii) To find P. I.

$$\text{P. I.} = \frac{1}{D^2 + 2D + 1} \, x \cos x$$

$$= \text{R. P. of } \frac{1}{D^2 + 2D + 1} e^{ix} \cdot x$$

$$= \text{R. P. of } e^{ix} \frac{1}{(D + i)^2 + 2(D + i) + 1} \, x$$

$$= \text{R. P. of } e^{ix} \frac{1}{D^2 + 2Di + i^2 + 2D + 2i + 1} \, x$$

$$= \text{R. P. of } e^{ix} \frac{1}{D^2 + 2Di + 2D + 2i} \, x$$

$$= \text{R. P. of } e^{ix} \frac{1}{2i\left(1 + \frac{D^2 + 2Di + 2D}{2i}\right)} \, x$$

$$= \text{R. P. of } e^{ix} \frac{1}{2i}\left(1 + \frac{D^2 + 2Di + 2D}{2i}\right)^{-1} x$$

$$= \frac{-1}{2} \, \text{R. P. of } e^{ix} i \left(1 - \frac{D^2 + 2Di + 2D}{2i} + \cdots \right) x$$

$$= \frac{-1}{2} \, \text{R. P. of } e^{ix} i \left(x + \frac{i}{2}(2i + 2)\right)$$

$$= \frac{-1}{2} \, \text{R. P. of } (\cos x + i \sin x)\left(ix - \frac{1}{2}(2i + 2)\right)$$

$$= \frac{-1}{2} \, \text{R. P. of } (\cos x + i \sin x)[(x - 1)i - 1]$$

$$= \frac{-1}{2}(-\cos x - (x-1)\sin x)$$

$$= \frac{1}{2}[\cos x + (x-1)\sin x]$$

iii) ∴ **C. S. is** $y = (C_1 + C_2 x)e^{-x} + \frac{1}{2}[\cos x + (x-1)\sin x]$

Example 54: Solve $\dfrac{d^2y}{dx^2} + 4y = x\ \sin x$

Solution: Given D. E. is in symbolic form $(D^2 + 4)y = x.\sin x$

  C. S. = C. F. + P. I.

i) To find C. F.

Its A. E. is $D^2 + 4 = 0$; $D^2 = -4$; $D = \pm 2i$

∴ C. F. = $C_1 \cos 2x + C_2 \sin 2x$

ii) To find P. I.

$$\text{P. I.} = \frac{1}{D^2 + 4}\ x\sin x$$

$$= \text{I. P. of }\ \frac{1}{D^2 + 4}\ e^{ix}\ x$$

$$= \text{I. P. of } e^{ix}\ \frac{1}{(D + i)^2 + 4}\ x$$

$$= \text{I. P. of } e^{ix}\ \frac{1}{D^2 + 2Di + 3}\ x \qquad\qquad \{\because i^2 = -1$$

$$= \text{I. P. of } e^{ix}\ \frac{1}{3\left(1 + \frac{D^2 + 2Di}{3}\right)}\ x$$

$$= \text{I. P. of } e^{ix}\ \frac{1}{3}\left(1 + \frac{D^2 + 2Di}{3}\right)^{-1}\ x$$

$$= \text{I. P. of } e^{ix}\ \frac{1}{3}\left(1 - \frac{D^2 + 2Di}{3} + \cdots\right)\ x$$

$$= \frac{1}{3}\ \text{I. P. of } e^{ix}\left(x - \frac{2}{3}i\right)$$

$$= \frac{1}{3}\ \text{I. P. of }(\cos x + i\sin x)\left(x - \frac{2}{3}i\right)$$

$$= \frac{1}{3}\left(-\frac{2}{3}\cos x + x\sin x\right)$$

iii) ∴ **C. S. is** $y = C_1 \cos 2x + C_2 \sin 2x + \dfrac{1}{3}\left(x\sin x - \dfrac{2}{3}\cos x\right)$

Example 55: Solve $\dfrac{d^4y}{dx^4} + \dfrac{2d^2y}{dx^2} + y = x^2 \cos x$

Solution: Given D. E. is in symbolic form as $(D^4 + 2D^2 + 1) y = x^2 \cos x$

C. S. = C. F. + P. I.

i) To find C. F.

Its A. E. is $(D^4 + 2D^2 + 1) = 0$; $(D^2 + 1)^2 = 0$; $\therefore D^2 = -1$; $D = \pm i, \ \pm i$

$\therefore$ C. F. = $(C_1 + C_2 x) \cos x + (C_3 + C_4 x) \sin x$

ii) To find P. I.

$$P.I. = \frac{1}{D^4 + 2D^2 + 1} x^2 \cos x$$

$$= \frac{1}{(D^2+1)^2} x^2 \cos x$$

$$= \text{R. P. of } \frac{1}{(D^2+1)^2} e^{ix}.x^2$$

$$= \text{R. P. Of } e^{ix} \frac{1}{[(D+i)^2+1]^2} x^2$$

$$= \text{R. P. of } e^{ix} \frac{1}{[D^2+2Di+i^2+1]^2} x^2$$

$$= \text{R. P. of } e^{ix} \frac{1}{(D^2+2Di)^2} x^2 \{\because i^2 = -1$$

$$= \text{R. P. of } e^{ix} \frac{1}{\left[2Di\left(1+\frac{D}{2i}\right)\right]^2} x^2$$

$$= \frac{-1}{4} \text{ R. P. of } e^{ix} \frac{1}{D^2}\left(1+\frac{D}{2i}\right)^{-2} x^2$$

$$= \frac{-1}{4} \text{ R. P. of } e^{ix} \frac{1}{D^2}\left(1 - 2\frac{D}{2i} + 3\left(\frac{D}{2i}\right)^2 - \cdots\right) x^2$$

$$= \frac{-1}{4} \text{ R. P. of } e^{ix} \frac{1}{D^2}\left(x^2 + iDx^2 + \frac{3}{-4}D^2x^2\right)$$

$$= \frac{-1}{4} \text{ R. P. of } e^{ix} \frac{1}{D^2}\left(x^2 + i2x - \frac{3}{4}2\right)$$

$$= \frac{-1}{4} \text{ R. P. of } e^{ix}\left(\frac{x^4}{12} + \frac{ix^3}{3} - \frac{3}{4}x^2\right)$$

$$= \frac{-1}{4} \text{ R. P. of } (\cos x + i\sin x)\left[\left(\frac{x^4}{12} - \frac{3x^2}{4}\right) + \frac{ix^3}{3}\right]$$

$$= \frac{-1}{4}\left[\left(\frac{x^4}{12} - \frac{3x^2}{4}\right)\cos x - \frac{x^3}{3}\sin x\right]$$

$$= \left(\frac{9x^2 - x^4}{48}\right) \cos x + \frac{x^3}{12} . \sin x$$

iii) $\therefore$ C. S. is $y = (c_1 + c_2 x) \cos x + (c_3 + c_4 x) \sin x + \dfrac{9x^2 - x^4}{48} . \cos x + \dfrac{x^3}{12} \sin x$

Example 56: Solve $\dfrac{d^2 y}{dx^2} - y = x^2 \sin 3x$

Solution: Given D. E. in symbolic form as $(D^2 - 1) y = x^2 \sin 3x$

   C. S. = C. F. + P. I.

i) To find C. F.

Its A. E. is $D^2 - 1 = 0$; $D^2 = 1$; $D = \pm 1$; $D = -1, \ 1$

$\therefore$ C. F. $= C_1 e^{-x} + C_2 e^x$

ii) To find P. I.

$$\text{P. I.} = \frac{1}{D^2 - 1} x^2 \sin 3x$$

$$= \text{I. P. of } \frac{1}{D^2 - 1} e^{3ix} . x^2$$

$$= \text{I. P. of } e^{3ix} \frac{1}{(D + 3i)^2 - 1} x^2$$

$$= \text{I. P. of } e^{3ix} \frac{1}{D^2 + 6Di + 9i^2 - 1} x^2$$

$$= \text{I. P. of } e^{3ix} \frac{1}{D^2 + 6Di - 10} x^2$$

$$= \text{I. P. of } e^{3ix} \frac{1}{-10\left(1 - \frac{D^2 + 6Di}{10}\right)} x^2$$

$$= \text{I. P. of } e^{3ix} \left(\frac{-1}{10}\right)\left(1 - \frac{D^2 + 6Di}{10}\right)^{-1} x^2$$

$$= \frac{-1}{10} \text{I. P. of } e^{3ix} \left[1 + \frac{D^2 + 6Di}{10} + \frac{36i^2 D^2}{100}\right] x^2$$

$$= \frac{-1}{10} \text{I. P. of } e^{3ix} \left[x^2 + \frac{1}{10}(2 + 12xi) - \frac{9}{25} . 2\right]$$

$$= \frac{-1}{10} \text{I. P. of } (\cos 3x + i \sin 3x) \left[x^2 + \frac{1}{5} + \frac{6}{5} xi - \frac{18}{25}\right]$$

$$= \frac{-1}{10} \text{I. P. of } (\cos 3x + i \sin 3x) \left[\left(x^2 - \frac{13}{25}\right) x^2 + i \frac{6x}{5}\right]$$

$$= \frac{-1}{10} \left[\frac{6}{5} x \cos 3x + \left(x^2 - \frac{13}{25}\right) \sin 3x\right]$$

iii) C.S. is $y = C_1 e^{-x} + C_2 e^x - \dfrac{1}{10}\left[\dfrac{6}{5} x \cos 3x + \left(x^2 - \dfrac{13}{25}\right)\sin 3x\right]$

Example 57: Solve $(D^2 + 4) y = x \sin^2 x$

Solution: Given D.E. is $(D^2 + 4) y = x \sin^2 x$

  C.S. = C.F. + P.I.

i) To find C. F.

Its A.E. is $D^2 + 4 = 0$; $D^2 = -4$; $D = \pm 2i$

∴ C.F. = $C_1 \cos 2x + C_2 \sin 2x$

ii) To find P.I.

$\begin{aligned}
\text{P.I} &= \frac{1}{D^2 + 4}\, x.\sin^2 x \\[2mm]
&= \frac{1}{D^2 + 4}\, x\left(\frac{1 - \cos 2x}{2}\right) \\[2mm]
&= \frac{1}{2}\left[\frac{1}{D^2 + 4}x - \frac{1}{D^2 + 4}x \cos 2x\right] \\[2mm]
&= \frac{1}{2}\left[\frac{1}{4\left(1 + \frac{D^2}{4}\right)}x - \text{R.P. of } \frac{1}{D^2 + 4}e^{2ix}\, x\right] \\[2mm]
&= \frac{1}{2}\left[\frac{1}{4}\left(1 + \frac{D^2}{4}\right)^{-1} x - \text{R.P. of } e^{2ix}\frac{1}{(D + 2i)^2 + 4}\, x\right] \\[2mm]
&= \frac{1}{2}\left[\frac{1}{4}\left(1 - \frac{D^2}{4} + \cdots\right)x - \text{R.P. of } e^{2ix}\frac{1}{D^2 + 4Di}\, x\right] \\[2mm]
&= \frac{1}{2}\left[\frac{1}{4}(x) - \text{R.P. of } e^{2ix}\frac{1}{D(D + 4i)}\, x\right] \\[2mm]
&= \frac{1}{2}\left[\frac{x}{4} - \text{R.P. of } e^{2ix}\frac{1}{D}\frac{1}{4i}\frac{1}{\left(1 + \frac{D}{4i}\right)}\, x\right] \\[2mm]
&= \frac{1}{2}\left[\frac{x}{4} - \text{R.P. of } e^{2ix}\frac{1}{D}\frac{1}{4i}\left(1 + \frac{D}{4i}\right)^{-1} x\right] \\[2mm]
&= \frac{1}{2}\left[\frac{x}{4} + \frac{1}{4}\, \text{R.P. of } e^{2ix}\frac{1}{D}\, i\left(1 - \frac{D}{4i} + \cdots\right)x\right] \\[2mm]
&= \frac{1}{2}\left[\frac{x}{4} + \frac{1}{4}\, \text{R.P. of } e^{2ix}\frac{1}{D}\left(xi - \frac{1}{4}\right)\right] \\[2mm]
&= \frac{1}{2}\left[\frac{x}{4} + \frac{1}{4}\, \text{R.P. of } (\cos 2x + i\sin 2x)\left(\frac{x^2}{2}i - \frac{x}{4}\right)\right]
\end{aligned}$

$$= \frac{1}{2}\left[\frac{x}{4} + \frac{1}{4}\left(-\frac{x}{4}\cos 2x - \frac{x^2}{2}\sin 2x\right)\right]$$

iii) ∴ C. S. is $y = C_1 \cos 2x + C_2 \sin 2x + \frac{1}{2}\left[\frac{x}{4} + \frac{1}{4}\left(-\frac{x}{4}\cos 2x - \frac{x^2}{2}\sin 2x\right)\right]$

Example 58: Solve $\dfrac{d^2y}{dx^2} + 3\dfrac{dy}{dx} + 2y = xe^{-x}\sin x$

Solution: Given D. E. in symbolic form $(D^2 + 3D + 2)\,y = xe^{-x}\sin x$

  C. S. = C. F. + P. I.

i) To find C. F.

Its A. E. is $(D^2 + 3D + 2) = 0$; $(D+1)(D+2) = 0$;   ∴ $D = -1, -2$

∴ C. F. $= C_1 e^{-x} + C_2 e^{-2x}$

ii) To find P. I.

$$P.I. = \frac{1}{D^2 + 3D + 2}\,xe^{-x}\sin x$$

$$= e^{-x}\frac{1}{(D-1)^2 + 3(D-1) + 2}\,x.\sin x$$

$$= e^{-x}\frac{1}{D^2 - 2D + 1 + 3D - 3 + 2}\,x.\sin x$$

$$= e^{-x}\frac{1}{D^2 + D}\,x.\sin x$$

$$= e^{-x}\,\text{I. P. of }\frac{1}{D^2 + D}\,e^{ix}x$$

$$= e^{-x}\,\text{I. P. of }e^{ix}\frac{1}{(D+i)^2 + (D+i)}\,x$$

$$= e^{-x}\,\text{I. P. of }e^{ix}\frac{1}{D^2 + 2Di - 1 + D + i}\,x$$

$$= e^{-x}\,\text{I. P. of }e^{ix}\frac{1}{D^2 + (1+2i)D + i - 1}\,x$$

$$= e^{-x}\,\text{I. P. of }e^{ix}\frac{1}{(i-1)\left(1 + \frac{D^2 + (1+2i)D}{i-1}\right)}\,x$$

$$= e^{-x}\,\text{I. P. of }e^{ix}\frac{i+1}{-2}\left(1 + \frac{D^2 + (1+2i)D}{i-1}\right)^{-1}x$$

$$= e^{-x}\,\text{I. P. of }e^{ix}\frac{i+1}{-2}\left(1 - \frac{D^2 + (1+2i)D}{i-1} + \cdots\right)x$$

$$= \frac{-1}{2}e^{-x}\,\text{I. P. of }(\cos x + i\sin x)(i+1)\left(x - \frac{1+2i}{i-1}\right)$$

$$= \frac{-1}{2} e^{-x} \text{ I. P. of } (\cos x + i \sin x)(i+1)\left(x - \frac{(1+2i)(i+1)}{-2}\right)$$

$$= \frac{-1}{2} e^{-x} \text{ I. P. of } (\cos x + i \sin x)\left[xi + \frac{i}{2}(i+1-2+2i) + x + \frac{1}{2}(i+1-2+2i)\right]$$

$$= \frac{-1}{2} e^{-x} \text{ I. P. of } (\cos x + i \sin x)\left[xi - \frac{3}{2} - \frac{i}{2} + x + \frac{3i}{2} - \frac{1}{2}\right]$$

$$= \frac{-1}{2} e^{-x} \text{ I. P. of } (\cos x + i \sin x)[(x+1)i + x - 2]$$

$$= \frac{-1}{2} e^{-x}[(x+1)\cos x + (x-2)\sin x]$$

iii) ∴ **C. S. is** $\quad y = C_1 e^{-x} + C_2 e^{-2x} - \dfrac{1}{2} e^{-x}[(x+1)\cos x + (x-2)\sin x]$

Example 59: Solve $(D^2 - 2D + 4)^2 y = x e^x \cos(\sqrt{3}\, x + \alpha)$

Solution: Given D. E. is $(D^2 - 2D + 4)^2 y = x e^x \cos(\sqrt{3}\, x + \alpha)$

C. S. = C. F. + P. I.

i) To find C. F.

Its A. E. is $(D^2 - 2D + 4)^2 = 0$;

$$D = \frac{2 \pm \sqrt{4-16}}{2}, \ \frac{2 \pm \sqrt{4-16}}{2} \ = \ \frac{2 \pm 2\sqrt{3}\, i}{2}, \ \frac{2 \pm 2\sqrt{3}\, i}{2}$$

∴ $D = 1 \pm \sqrt{3}\, i, \ 1 \pm \sqrt{3}\, i$

∴ C. P. $= e^x\left[(C_1 + C_2 x)\cos\sqrt{3}\, x + (C_3 + C_4 x)\sin\sqrt{3}\, x\right]$

ii) To find P. I.

$$\text{P. I.} \ = \frac{1}{(D^2 - 2D + 4)^2} x e^x \cos(\sqrt{3}\, x + \alpha)$$

$$= e^x \frac{1}{[(D+1)^2 - 2(D+1)+4]^2} x \cos(\sqrt{3}\, x + \alpha)$$

$$= e^x \frac{1}{[D^2 + 2D + 1 - 2D - 2 + 4]^2} x . \cos(\sqrt{3}\, x + \alpha)$$

$$= e^x \frac{1}{[D^2 + 3]^2} x . \cos(\sqrt{3}\, x + \alpha)$$

$$= e^x \text{ R. P. of } \frac{1}{[D^2 + 3]^2} e^{i(x\sqrt{3}+\alpha)} . x$$

$$= e^x \text{ R. P. of } e^{i(x\sqrt{3}+\alpha)} \frac{1}{\left[(D+i\sqrt{3})^2 + 3\right]^2} . x$$

$$= e^x . \text{R. P. of } e^{i(x\sqrt{3}+\alpha)} \frac{1}{\left[D + 2\sqrt{3}\, D\, i + 3i^2 + 3\right]^2} . x$$

$$= e^x \text{ R. P. of } e^{i(x\sqrt{3}+\alpha)} \frac{1}{\left[2\sqrt{3}Di\left(1+\frac{D}{2\sqrt{3}i}\right)\right]^2} x$$

$$= e^x \text{ R. P. of } e^{i(x\sqrt{3}+\alpha)} \frac{1}{-12D^2}\left(1+\frac{D}{2i\sqrt{3}}\right)^{-2} x$$

$$= \frac{-e^x}{12} \text{ R. P. of } e^{i(x\sqrt{3}+\alpha)} \frac{1}{D^2}\left(1-2\frac{D}{2i\sqrt{3}}+\cdots\right)x$$

$$= \frac{-e^x}{12} \text{ R. P. of } e^{i(x\sqrt{3}+\alpha)} \frac{1}{D^2}\left(x+i\frac{1}{\sqrt{3}}\right)$$

$$= \frac{-e^x}{12} \text{ R. P. of } \left[\cos(x\sqrt{3}+\alpha)+i\sin(x\sqrt{3}+\alpha)\right]\left[\frac{x^3}{6}+i\frac{1}{\sqrt{3}}\frac{x^2}{2}\right]$$

$$= \frac{-e^x}{12}\left[\frac{x^3}{6}\cos(x\sqrt{3}+\alpha)-\frac{x^2}{2\sqrt{3}}\sin(x\sqrt{3}+\alpha)\right]$$

iii) $\therefore$ C.S. is $y = e^x\left[(C_1+C_2x)\cos\sqrt{3}x+(C_3+C_4x)\sin\sqrt{3}x\right]$

$$-\frac{e^x}{12}\left[\frac{x^3}{6}\cos(x\sqrt{3}+\alpha)-\frac{x^2}{2\sqrt{3}}\sin(x\sqrt{3}+\alpha)\right]$$

Example 60: Solve $\dfrac{d^2y}{dx^2} - 4y = x\sin h\,x$

Solution: Given D. E. in symbolic form as $(D^2-4)\,y = x\sinh x$

C. S. = C. F. + P. I.

i) To find C. F.

Its A. E. is $D^2-4 = 0; D^2 = 4; \quad D = \pm 2$

$\therefore$ C. F. = $C_1 e^{-2x} + C_2 e^{2x}$

ii) To find P. I.

$$\text{P. I.} = \frac{1}{D^2-4} x \sinh x$$

$$= \frac{1}{D^2-4}x\left(\frac{e^x-e^{-x}}{2}\right)$$

$$= \frac{1}{2}\left[\frac{1}{D^2-4}xe^x - \frac{1}{D^2-4}xe^{-x}\right]$$

$$= \frac{1}{2}\left[e^x\frac{1}{(D+1)^2-4}x - e^{-x}\frac{1}{(D-1)^2-4}x\right]$$

$$= \frac{1}{2}\left[e^x\frac{1}{D^2+2D-3}x - e^{-x}\frac{1}{D^2-2D-3}x\right]$$

$$= \frac{1}{2}\left[e^x\frac{1}{-3\left(1-\frac{D^2+2D}{3}\right)}x - e^{-x}\frac{1}{-3\left(1-\frac{D^2-2D}{3}\right)}x\right]$$

$$= \frac{1}{2}\left[\frac{e^x}{-3}\left(1 - \frac{D^2 + 2D}{3}\right)^{-1} x + \frac{e^{-x}}{3}\left(1 - \frac{D^2 - 2D}{3}\right)^{-1} x\right]$$

$$= \frac{1}{2}\left[\frac{e^x}{-3}\left(1 + \frac{D^2 + 2D}{3} + \dots\right)x + \frac{e^{-x}}{3}\left(1 + \frac{D^2 - 2D}{3} + \dots\right)\right]x$$

$$= \frac{1}{2}\left[\frac{e^x}{-3}\left(x + \frac{2}{3}\right) + \frac{e^{-x}}{3}\left(x - \frac{2}{3}\right)\right]$$

$$= \frac{1}{2}\left[\frac{e^x \, x}{-3} - \frac{e^x \, 2}{9} + \frac{e^{-x} \, x}{3} - \frac{e^{-x} \, 2}{3}\right]$$

$$= \frac{-1}{6}\left[(e^x - e^{-x})x + \frac{2}{3}(e^x + e^{-x})\right]$$

$$= \frac{-1}{6}\left[2 \sinh x \, (x) + \frac{2}{3} 2 \cosh x\right]$$

$$= -\frac{x}{3} \sinh x - \frac{2}{9} \cosh x$$

iii) ∴ C. S. is $y = C_1 e^{-2x} + C_2 e^{2x} - \frac{x}{3} \sinh x - \frac{2}{9} \cosh x$

Example 61: Solve $(D^3 + 3D) y = \cosh 2x . \sinh 3x$

Solution: Given D. E. is $(D^3 + 3D) y = \cosh 2x . \sinh 3x$

C. S. = C. F. + P. I.

i) **To find C. F.**

Its A. E. is $D^2 + 3D = 0$; $D(D^2 + 3) = 0$

∴ $D = 0$, $D^2 = -3$; $D = 0$, $\pm \sqrt{3}i$

∴ C. F. = $C_1 + C_2 \cos \sqrt{3} \, x + C_3 \sin \sqrt{3} \, x$

ii) **To find P. I.**

$$\text{P. I.} = \frac{1}{D^3 + 3D} \cosh 2x . \sinh 3x$$

$$= \frac{1}{D . D^2 + 3D} \frac{1}{2} (\sinh 5x + \sin hx)$$

$$\left\{\begin{array}{l} \because \cosh A . \sinh B = \frac{1}{2}[\sinh(A + B) + \sinh(A - B)] \\ \because \sinh(-x) = \sinh x \; ; \cosh(-x) = \cosh x \end{array}\right.$$

$$= \frac{1}{2}\left[\frac{1}{D \, D^2 + 3D} \sinh 5x + \frac{1}{D \, D^2 + 3D} \sinh x\right]$$

$$= \frac{1}{2}\left[\frac{1}{25D + 3D} \sinh 5x + \frac{1}{D + 3D} \sinh x\right]$$

$$= \frac{1}{2}\left[\frac{1}{28} \frac{1}{D} \sinh 5x + \frac{1}{4} \frac{1}{D} \sinh x\right]$$

$$= \frac{1}{56} \frac{\cosh 5x}{5} + \frac{1}{8} \cosh x \left\{ \because \int \sinh \theta \ d\theta = \cosh \theta \right.$$

$$= \frac{1}{280} \cosh 5x + \frac{1}{8} \cosh x$$

iii) ∴ C.S. is $y = C_1 + C_2 \cos \sqrt{3}\, x + C_3 \sin \sqrt{3}\, x + \dfrac{1}{280} \cosh 5x + \dfrac{1}{8} \cosh x$

Example 62: Solve $(D^2 - 1)y = \cosh x . \cos x$

Solution: $(D^2 - 1)\, y = \cosh x . \cos x$

  C.S. = C.F. + P.I.

i) To find C.F.

Its A.E. is $D^2 - 1 = 0$; $D^2 = 1$; $D = \pm 1$; $D = 1, \ -1$

∴ C.F. $= C_1 e^x + C_2 e^{-x}$

ii) To find P.I.

$$\text{P.I.} \ = \frac{1}{D^2 - 1} \cos hx \ . \cos x.$$

$$= \frac{1}{D^2 - 1} \left(\frac{e^x + e^{-x}}{2} \right) \cos x \left\{ \because \cos h\theta = \frac{e^\theta + e^{-\theta}}{2} \right.$$

$$= \frac{1}{2} \left[\frac{1}{D^2 - 1} e^x \cos x + \frac{1}{D^2 - 1} e^{-x} \cos x \right]$$

$$= \frac{1}{2} \left[e^x \frac{1}{(D + 1)^2 - 1} \cos x + e^{-x} \frac{1}{(D - 1)^2 - 1} \cos x \right]$$

$$= \frac{1}{2} \left[e^x \frac{1}{D^2 + 2D + 1 - 1} \cos x + e^{-x} \frac{1}{D^2 - 2D \pm 1} \cos x \right]$$

$$= \frac{1}{2} \left[e^x \frac{1}{D^2 + 2D} \cos x + e^{-x} \frac{1}{D^2 - 2D} \cos x \right]$$

$$= \frac{1}{2} \left[e^x \frac{1}{(-1)(1)^2 + 2D} \cos x + e^{-x} \frac{1}{(-1)(1)^2 - 2D} \cos x \right]$$

$$= \frac{1}{2} \left[e^x \frac{1}{2D - 1} \cos x + e^{-x} \frac{1}{-2D - 1} \cos x \right]$$

$$= \frac{1}{2} \left[e^x \frac{2D + 1}{4D^2 - 1} \cos x + e^{-x} \frac{-2D + 1}{4D^2 - 1} . \cos x \right]$$

$$= \frac{1}{2} \left[e^x \frac{(2D + 1).\cos x}{4(-1)(1)^2 - 1} + e^{-x} \frac{(-2D + 1).\cos x}{4(-1)(1)^2 - 1} \right]$$

$$= \frac{1}{2} \left[\frac{e^x}{-5} [2(-\sin x) + \cos x] + \frac{e^{-x}}{-5} (2\sin x + \cos x) \right]$$

$$= \frac{e^x}{5} \sin x - \frac{e^x}{10} \cos x - \frac{e^{-x}}{5} \sin x - \frac{e^{-x}}{10} \cos x$$

$$= \frac{1}{5}(e^x - e^{-x})\sin x - \frac{1}{10}(e^x + e^{-x})\cos x$$

$$= \frac{1}{5}2\sinh x . \sin x - \frac{1}{10}2\cosh x . \cos x$$

iii) $\therefore$ C. S. is $y = C_1 e^x + C_2 e^{-x} + \frac{1}{5}(2\sinh x . \sin x - \cosh x . \cos x)$

7   Equations Reducible to linear equations with constant coefficients

Now we shall study two forms of linear different equation with variable coefficients which can be reduced to linear differential equarions with constant coefficient by suitable substitutions.

     i.   **Cauchy's homogeneous linear equation.**

     ii.   **Legendre's linear equation.**

7.i   Cauchy's homogeneous linear equation

An equation of the form:

$$x^n \frac{d^n y}{dx^n} + k_1 x^{n-1} \frac{d^{n-1} y}{dx^{n-1}} + \ ... \ + k_{n-1} x \frac{dy}{dx} + k_n y = X \qquad ...\,...\,(1)$$

Where X is a function of x and $k_1, k_2 \,...\, k_{n-1}$ are constants is called Cauchy's homogenous linear equation.

Such equations can be reduced to linear differential equations with constant coefficient, by putting

Put $x = e^z$ or $z = \log x$

     Let $D = \dfrac{d}{dz}$

     i) $x\dfrac{dy}{dx} \quad = Dy$

     ii) $x^2 \dfrac{d^2 y}{dx^2} = D(D-1)\,y$

     iii) $x^3 \dfrac{d^3 y}{dx^3} = D(D-1)(D-2)y$ etc.

After making these substitutions in equation (1), there results a linear equation with constant coefficient which can be solved as before.

7.i.a   Examples on Cauchy's homogeneous linear equation

Example 63: Solve $x^2 \dfrac{d^2 y}{dx^2} - x \dfrac{dy}{dx} + y = \log x$

Solution: Given D. E. $x^2 \dfrac{d^2 y}{dx^2} - x\dfrac{dy}{dx} + y = \log x$ $...\,...\,(1)$

This is a Cauchy's homogeneous linear equation

It can be reduced to LDE with constant coefficient by putting

$\therefore$ Put $x = e^z$, $z = \log x$,   Let $D = \dfrac{d}{dz}$

$\therefore x^2 \dfrac{d^2 y}{dx^2} = D(D-1)y$; $x \dfrac{dy}{dx} = Dy$

$\therefore$ Equation (1) becomes, $D(D-1)y - Dy + y = z$

i.e. $[D(D-1) - D + 1]y = z$

$(D^2 - D - D + 1)y = z$

$(D^2 - 2D + 1)y = z$ ... Which is linear D.E. with constant coefficient

$\therefore$ C.S.= C.F.+P.I.

i) **To find C.F.**

Its A.E. is $D^2 - 2D + 1 = 0$; $(D-1)^2 = 0$; $D = 1, 1$

$\therefore$ C.F.= $(C_1 + C_2 z)e^z = (C_1 + C_2 \log x)x$

ii) **To find P.I.**

$\therefore$ P.I.$= \dfrac{1}{(D^2 - 2D + 1)}\, z = \dfrac{1}{(D-1)^2}z = \dfrac{1}{(1-D)^2}z$

$= (1-D)^{-2}z = (1 + 2D + 3D^2 + \ldots z)$

$= z + 2 = \log x + 2$

iii) $\therefore$ **C.S. is** $y = (C_1 + C_2 \log x)x + \log x + 2$

Example 64: Solve $x^2 \dfrac{d^2 y}{dx^2} + 3x \dfrac{dy}{dx} + y = \dfrac{1}{(1-x)^2}$

Solution: Given D.E. is, $x^2 \dfrac{d^2 y}{dx^2} + 3x \dfrac{dy}{dx} + y = \dfrac{1}{(1-x)^2}$ ...... (1)

This is a Cauchy's homogeneous linear equation

$\therefore$ Put $z = \log x$, $x = e^z$,   Let $D = \dfrac{d}{dz}$

$\therefore x^2 \dfrac{d^2 y}{dx^2} = D(D-1)\, y$; $x \dfrac{dy}{dx} = Dy$

$\therefore$ Equation (1) becomes, $[D(D-1) + 3D + 1]y = \dfrac{1}{(1-e^z)^2}$

i.e. $(D^2 + 2D + 1)y = \dfrac{1}{(1-e^z)^2}$ ... Which is linear D.E. with constant coefficient

$\therefore$ C.S.= C.F.+ P.I.

i) **To find C.F.**

Its A.E. is $(D^2 + 2D + 1) = 0$; $(D+1)^2 = 0$; $D = -1, -1$

$$\therefore \text{ C.F.} = (C_1 + C_2 z)e^{-z} = (C_1 + C_2 \log x)\frac{1}{x}$$

ii) To find P. I.

$$\text{P. I.} = \frac{1}{D^2 + 2D + 1}\frac{1}{(1 - e^z)^2} = \frac{1}{(D + 1)^2}(1 - e^z)^{-2}$$

$$\text{P. I.} = \frac{1}{(D + 1)}\left[\frac{1}{(D + 1)}(1 - e^z)^{-2}\right]$$

$$\text{P. I.} = \frac{1}{(D + 1)} \cdot u \qquad\qquad \dots\dots (2)$$

$$\text{Where } u = \frac{1}{D + 1}(1 - e^z)^{-2}$$

$$\text{Consider } u = \frac{1}{D + 1}(1 - e^z)^{-2}$$

$$(D + 1)u = (1 - e^z)^{-2}$$

$$\frac{du}{dz} + u = (1 - e^z)^{-2}$$

$$\text{Which is Leibnitz's linear equation}\frac{dy}{dx} + Py = Q$$

$$\left\{ \begin{array}{l} \therefore \text{ I. F.} = e^{\int Pdx}, \quad \text{where} \quad P = 1, \ Q = (1 - e^z)^{-2} \\ \text{And It's solution is,} \quad y\,(\text{I. F.}) = \int Q.(\text{I. F.})\,dx \end{array} \right.$$

$$\therefore \text{I. F.} = e^{\int Pdz} = e^{\int 1dz} = e^z$$

$$\therefore \text{Its solution is } ue^z = \int \frac{1}{(1 - e^z)^2} \cdot e^z dz \left\{ \begin{array}{l} \because \text{Put } 1 - e^z = t \\ -e^z dz = dt; \ e^z dz = -dt \end{array} \right.$$

$$= \int \frac{-dt}{t^2} = -\int t^{-2}dt = \frac{-t^{-2+1}}{-2 + 1} = \frac{-t^{-1}}{-1} = \frac{1}{t} = \frac{1}{1 - e^z}$$

$$\therefore \quad ue^z = \frac{1}{1 - e^z}$$

$$\therefore \quad u = \frac{e^{-z}}{1 - e^z}$$

$$\therefore \text{Equation (2) becomes,}$$

$$\text{P. I.} = \frac{1}{D + 1}\frac{e^{-z}}{1 - z}$$

$$= e^{-z}\int e^z \frac{e^{-z}}{1 - e^z}\,dz$$

$$= e^z \int \frac{1}{1 - e^z}\,dz$$

$$= e^{-z} \int \frac{e^{-z}}{e^{-z} - 1} dz \qquad \{\because \text{Multiply and dividing } N^r \& D^r \text{ by } e^{-z}$$

$$= e^{-z} \int \frac{e^{-z}}{e^{-z} - 1} dz$$

$$= e^{-z} \int \frac{-d\theta}{\theta} \left\{ \begin{array}{l} \text{put } e^{-z} - 1 = \theta \\ -e^{-z}.dz = d\theta; \quad e^{-z}.dz = -d\theta \end{array} \right.$$

$$= e^{-z} [-\log \theta]$$

$$= e^{-z} [-\log(e^{-z} - 1)]$$

$$= -e^{-z} \log(e^{-z} - 1)$$

$$= -\frac{1}{x} \log \left(\frac{1}{x} - 1 \right) \quad = -\frac{1}{x} \log \left(\frac{1-x}{x} \right)$$

$$= \frac{1}{x} \log \left(\frac{x}{1-x} \right)$$

iii) $\therefore$ C. S. is $\quad y = \left[C_1 + C_2 \log x + \log \left(\frac{x}{1-x} \right) \right] \frac{1}{x}$

Example 65: Solve $x^2 \dfrac{d^2y}{dx^2} + x \dfrac{dy}{dx} + y = \log x \sin(\log x)$

Solution: Given D. E. is $\quad x^2 \dfrac{d^2y}{dx^2} + x \dfrac{dy}{dx} + y = \log x \sin(\log x) \qquad \ldots \ldots (1)$

This is Cauchy's LDE. It can be reduced to LDE with constant coefficient by putting

$$x = e^z, \qquad z = \log x, \qquad \text{Let} \quad D = \frac{d}{dz}$$

$$\therefore x^2 \frac{d^2y}{dx^2} = D(D-1)y \quad ; \quad x \frac{dy}{dx} = Dy$$

$\therefore$ Equation $(1) \rightarrow [D(D-1) + D + 1]y = z . \sin z$

$\therefore \left(D^2 + 1 \right) y = z . \sin z \qquad \ldots$ Which is LDE with constant coefficient

$\therefore$ C. S. = C. F. + P. I.

i) To find C. F.

Its A. E. is $D^2 + 1 = 0$; $D^2 = -1$; $D = \pm i$

$\therefore$ C. F. = $C_1 \cos z + C_2 \sin z = C_1 \cos(\log x) + C_2 \sin(\log x)$

ii) To find P. I.

$$\text{P. I.} = \frac{1}{D^2 + 1} z \sin z$$

$$= \text{I. P. of } \frac{1}{D^2 + 1} z \, e^{iz}$$

$$= \text{I. P. of } e^{iz} \frac{1}{(D+i)^2 + 1} z$$

$$= \text{I.P. of } e^{iz} \frac{1}{D^2 + 2Di} \, z$$

$$= \text{I.P. of } e^{iz} \frac{1}{D} \frac{1}{(D + 2i)} \, z$$

$$= \text{I.P. of } e^{iz} \frac{1}{D} \frac{1}{2i\left(1 + \frac{D}{2i}\right)} \, z$$

$$= \text{I.P. of } e^{iz} \frac{1}{2}(-i)\frac{1}{D}\left(1 + \frac{D}{2i}\right)^{-1} z$$

$$= \frac{1}{2}\text{I.P. of } e^{iz}(-i)\frac{1}{D}\left(1 - \frac{D}{2i} + \cdots\right)z$$

$$= \frac{1}{2} \text{ I.P. of } (\cos z + i \sin z)(-i)\frac{1}{D}\left(z - \frac{1}{2i}\right)$$

$$= \frac{1}{2} \text{ I.P. of } (\cos z + i \sin z)(-i)\left(\frac{z^2}{2} + \frac{i}{2}z\right)$$

$$= \frac{1}{2} \text{ I.P. of } (\cos z + i \sin z)\left(\frac{-z^2 i}{2} + \frac{1}{2}z\right)$$

$$= \frac{1}{2}\left[\frac{-z^2}{2}\cos z + \frac{z}{2}\sin z\right] \quad = \frac{-z^2}{4}\cos z + \frac{z}{4}\sin z$$

$$= \frac{-1}{4}(\log x)^2 \cos(\log x) + \frac{\log x}{4}\sin(\log x)$$

iii) $\therefore$ C. S. is $y = C_1 \cos(\log x) + C_2 \sin(\log x) - \frac{1}{4}(\log x)^2 \cos(\log x)$

$$+ \frac{1}{4}(\log x).\sin(\log x)$$

Example 66: Solve $x^2 \dfrac{d^2y}{dx^2} - 3x \dfrac{dy}{dx} + y = \log x \dfrac{\sin(\log x) + 1}{x}$

Solution: Given D. E. is $x^2 \dfrac{d^2y}{dx} - 3x\dfrac{dy}{dx} + y = \log x.\dfrac{\sin(\log x) + 1}{x}$ ... ... (1)

This is Cauchy's D. E., $\therefore$ Put $x = e^z$, $z = \log x$, $\dfrac{d}{dz} = D$

$x^2 \dfrac{d^2y}{dx^2} = D(D - 1)y$; $x\dfrac{dy}{dx} = Dy$

$\therefore$ Equation $(1) \rightarrow [D(D - 1) - 3D + 1]y = z\dfrac{\sin(z) + 1}{e^z}$

$\therefore \left(D^2 - 4D + 1\right)y = z.\,e^{-z}[\sin(z) + 1]$ ... Which is L. D. E. with constant coefficient

$\therefore$ C. S. $=$ C. F. $+$ P. I.

i) **To find C. F.**

Its A. E. is $D^2 - 4D + 1 = 0$; $D = 2 \pm \sqrt{3} = 2 + \sqrt{3}, \ 2 - \sqrt{3}$

$\therefore$ C. F. $= C_1 e^{(2+\sqrt{3})z} + C_2 e^{(2-\sqrt{3})z} = e^{2z}\left(C_1 e^{\sqrt{3}z} + C_2 e^{-\sqrt{3}z}\right)$

$\therefore$ C. F. $= x^2 \left(C_1 x^{\sqrt{3}} + C_2 x^{-\sqrt{3}}\right)$

ii) To find P. I.

$$\text{P.I.} = \frac{1}{D^2 - 4D + 1} z \, e^{-z}(\sin z + 1)$$

$$= e^{-z} \frac{1}{(D-1)^2 - 4(D-1) + 1} z\,(\sin z + 1)$$

$$= e^{-z} \frac{1}{D^2 - 2D + 1 - 4D + 4 + 1} z\,(\sin z + 1)$$

$$= e^{-z} \left[\frac{1}{D^2 - 6D + 6} z.\sin z + \frac{1}{D^2 - 6D + 6}z\right]$$

$$= e^{-z} \left[\text{I.P. of } \frac{1}{D^2 - 6D + 6} e^{iz}.z + \frac{1}{6\left(1 + \frac{D^2-6D}{6}\right)}z\right]$$

$$= e^{-z} \left[\text{I.P. of } e^{iz}\frac{1}{(D+i)^2 - 6(D+i) + 6}z + \frac{1}{6}\left(1 + \frac{D^2 - 6D}{6}\right)^{-1}z\right]$$

$$= e^{-z} \left[\text{I.P. of } e^{iz}\frac{1}{D^2 + 2Di - 6D - 6i + 5}z + \frac{1}{6}\left(1 - \frac{D^2 - 6D}{6} + \cdots\right)z\right]$$

$$= e^{-z} \left[\text{I.P. of } e^{iz}\frac{1}{(5-6i)\left(1 + \frac{D^2+2Di-6D}{(5-6i)}\right)}z + \frac{1}{6}\left(z - \frac{1}{6}(-6)\right)\right]$$

$$= e^{-z} \left[\text{I.P. of } \frac{e^{iz}}{5-6i}\left(1 + \frac{D^2 + 2Di - 6D}{5-6i}\right)^{-1}z + \frac{1}{6}z + \frac{1}{6}\right]$$

$$= e^{-z} \left[\text{I.P. of}\frac{e^{iz}(5+6i)}{(5^2-(6i)^2)}\left(1 - \frac{D^2 + 2Di - 6D}{5-6i} + \cdots\right)z + \frac{1}{6}(z+1)\right]$$

$$= e^{-z} \left[\frac{1}{61}\text{ I.P. of } (\cos z + i\sin z)(5+6i)\left(z - \frac{(5+6i)(2i-6)}{61}\right) + \frac{1}{6}(z+1)\right]$$

$$= e^{-z} \left[\frac{1}{61}\text{ I.P. of } (\cos z + i\sin z)(5+6i)\left(z - \frac{1}{61}(-26i - 42)\right) + \frac{1}{6}(z+1)\right]$$

$$= e^{-z} \left[\frac{1}{61}\text{I.P. of } (\cos z + i\sin z)\left(5z + \frac{5}{61}(26i + 42) + 6zi + \frac{6i(26i + 42)}{61}\right)\right.$$
$$\left. + \frac{1}{6}(z+1)\right]$$

$$= e^{-z}\left[\frac{1}{61}\left(\frac{26.5}{61} + 6z + \frac{42.6}{61}\right)\cos z + \frac{1}{61}\left(5z + \frac{42.5}{61} - \frac{26.6}{61}\right)\sin z + \frac{1}{6}(z+1)\right]$$

$$= e^{-z}\left[\frac{1}{3721}(382 + 366z)\cos z + \frac{1}{3721}(305z + 54)\sin z + \frac{1}{6}(z+1)\right]$$

$$= \frac{1}{x}\left[\frac{1}{3721}(382 + 366\log x)\cos(\log x) + \frac{1}{3721}(305\log x + 54)\sin(\log x)\right.$$
$$\left. + \frac{1}{6}(\log x + 1)\right]$$

iii) C.S. is

$$y = x^2\left(C_1 x^{\sqrt{3}} + C_2 x^{-\sqrt{3}}\right) + \frac{1}{x}\left[\frac{1}{3721}(382 + 366\log x)\cos(\log x)\right.$$

$$\left. + \frac{1}{3721}(305\log x + 54)\sin(\log x) + \frac{1}{6}(\log x + 1)\right]$$

Example 67: Solve $x^2\dfrac{d^2y}{dx^2} + 4x\dfrac{dy}{dx} + 2y = e^x$

Solution: Given D.E. is $x^2\dfrac{d^2y}{dx^2} + 4x\dfrac{dy}{dx} + 2y = e^x$ ...... (1)

This is Cauchy's D.E.,   ∴ Put $x = e^z$, $z = \log x$, $D = \dfrac{d}{dz}$

$$x^2\frac{d^2y}{dx^2} = D(D-1)y \quad ; \quad x\frac{dy}{dx} = Dy$$

∴ Equation(1)→$[D(D-1) + 4D + 2]\,y = e^{e^z}$

$$\left(D^2 + 3D + 2\right)y = e^{e^z} \qquad \text{... Which is L.D.E. with constant coefficient}$$

∴ C.S. = C.F. + P.I.

i) To find C.F.

Its A.E. is $D^2 + 3D + 2 = 0$; $(D+1)(D+2) = 0$;   ∴ $D = -1, -2$

∴ C.F. = $C_1 e^{-z} + C_2 e^{-2z} = C_1 x^{-1} + C_2 x^{-2}$

ii) To find P.I.

$$\therefore \text{P.I.} = \frac{1}{D^2 + 3D + 2}e^{e^z} = \frac{1}{(D+1)(D+2)}e^{e^z}$$

$$= \left[\frac{1}{D+1} - \frac{1}{D+2}\right]e^{e^z} \qquad\qquad \{\because \text{By partial fraction}$$

$$= \frac{1}{D+1}e^{e^z} - \frac{1}{D+2}e^{e^z}$$

$$= \frac{1}{D+1}e^{-z}e^{e^z} - \frac{1}{D+2}e^{-2z}e^{2z}e^{e^z} \qquad\qquad \text{...Note} \qquad \{\because \text{Adjustment}$$

$$= e^{-z}\frac{1}{(D-1)+1}e^z e^{e^z} - e^{-2z}\frac{1}{(D-2)+2}e^{2z}.e^{e^z}$$

$$= e^{-z} \frac{1}{D} e^z e^{e^z} - e^{-2z} \frac{1}{D} e^{2z}. e^{e^z}$$

$$= e^{-z} \int e^z e^{e^z} dz - e^{-2z} \int e^{2z}. e^{e^z} dz$$

$$= e^{-z} \int e^t dt - e^{-2z} \int t e^t dt \begin{cases} \because \text{ put } e^z = t \\ e^z dz = dt \end{cases}$$

$$= e^{-z} e^{e^z} - e^{-2z} (t e^t - e^t)$$

$$= e^{-z} e^{e^z} - e^{-2z} (e^z e^{e^z} - e^{e^z})$$

$$= x^{-1} e^x - x^{-2} (x e^x - e^x)$$

$$= x^{-1} e^x - x^{-1} e^x + x^{-2} e^x$$

$$= x^{-2} e^x$$

iii) ∴ **C. S. is** $y = C_1 x^{-1} + C_2 x^{-2} + x^{-2} e^x$

Example 68: Solve $(x^3 D^3 + x^2 D^2 - 2)y = x + \dfrac{1}{x^3}$

Solution: Given D. E. is, $(x^3 D^3 + x^2 D^2 - 2)y = x + \dfrac{1}{x^3}$ ... ... (1)

This is Cauchy's D. E.,   ∴ Put $x = e^z$, $z = \log x$; $\theta = \dfrac{d}{dz}$

∴ $x^3 D^3 = \theta (\theta - 1)(\theta - 2)y$;

$x^2 D^2 = \theta (\theta - 1)y$

∴ Equation (1) → $[\theta(\theta - 1)(\theta - 2) + \theta(\theta - 1) - 2]y = e^z + \dfrac{1}{e^{3z}}$

$[\theta^3 - \theta^2 - 2\theta^2 + 2\theta + \theta^2 - \theta - 2]y = e^z + e^{-3z}$

∴ $(\theta^3 - 2\theta^2 + \theta - 2)y = (e^z + e^{-3z})$ ... Which is L. D. E. with constant coefficient

∴ C. S. = C. F. + P. I.

i) **To find C. F.**

Its A. E. is $\theta^3 - 2\theta^2 + \theta - 2 = 0$;   ∴ $\theta = 2, \pm i$

∴ C. F. = $C_1 e^{2z} + C_2 \cos z + C_3 \sin z = C_1 x^2 + C_2 \cos(\log x) + C_3 \sin(\log x)$

ii) **To find P. I.**

∴ P. I. $= \dfrac{1}{\theta^3 - 2\theta^2 + \theta - 2} (e^z + e^{-3z})$

$= \dfrac{1}{\theta^3 - 2\theta^2 + \theta - 2} e^z + \dfrac{1}{\theta^3 - 2\theta^2 + \theta - 2} e^{-3z}$

$= \dfrac{e^z}{1 - 2(1) + 1 - 2} + \dfrac{e^{-3z}}{(-3)^3 - 2(-3)^2 + (-3) - 2}$

$= \dfrac{e^z}{-2} + \dfrac{e^{-3z}}{-50} = -\dfrac{1}{2} e^z - \dfrac{1}{50} e^{-3z}$

$$= -\frac{x}{2} - \frac{x^{-3}}{50}$$

iii) ∴ C.S. is $y = C_1 x^2 + C_2 \cos(\log x) + C_3 \sin(\log x) - \dfrac{x}{2} - \dfrac{x^{-3}}{50}$

Example 69: Evaluate $x^3 \dfrac{d^3 y}{dx^3} + 3x^2 \dfrac{d^2 y}{dx^2} + x\dfrac{dy}{dx} - y = 3x - 7$

Solution: Given D. E. is $x^3 \dfrac{d^3 y}{dx^3} + 3x^2 \dfrac{d^2 y}{dx^2} + x\dfrac{dy}{dx} - y = 3x - 7$ ... ... (1)

This is Cauchy's D. E.,   ∴ Put $x = e^z$, $z = \log x$, $\dfrac{d}{dz} = D$

∴ $x^3 \dfrac{d^3 y}{dx^3} = D(D-1)(D-2)y$; $x^2 \dfrac{d^2 y}{dx^2} = D(D-1)y$; $x\dfrac{dy}{dx} = Dy$

∴ Equation(1)→$[D(D-1)(D-2) + 3D(D-1) + D - 1]y = 3e^z - 7$

∴ $(D^3 - 2D^2 - D^2 + 2D + 3D^2 - 3D + D - 1)y = 3e^z - 7$

∴ $(D^3 - 1)y = 3e^z - 7$ ... Which is L. D. E. with constant coefficient

∴ C. S. = C. F. + P. I.

i) To find C. F.

∴ Its A. E. is $D^3 - 1 = 0$; $(D-1)(D^2 + D + 1) = 0$

∴ $D = 1, \dfrac{-1}{2} \pm i\dfrac{\sqrt{3}}{2}$

∴ C. F. = $C_1 e^z + e^{-\frac{1}{2}z}\left(C_2 \cos\dfrac{\sqrt{3}}{2}z + C_3 \sin\dfrac{\sqrt{3}}{2}z \right)$

∴ C. F. = $C_1 x + x^{-\frac{1}{2}}\left[C_2 \cos\dfrac{\sqrt{3}}{2}(\log x) + C_3 \sin\dfrac{\sqrt{3}}{2}(\log x) \right]$

ii) To find P. I.

P. I. $= \dfrac{1}{D^3 - 1}(3e^z - 7)$

$= 3\dfrac{1}{D^3 - 1}e^z - 7\dfrac{1}{D^3 - 1}e^z$

$= 3\dfrac{z}{3D^2}e^z - 7\dfrac{1}{0^3 - 1}$

$= 3\dfrac{ze^z}{3(1)^2} - 7\dfrac{1}{-1}$ $= ze^z + 7$

$= \log x\,(x) + 7$

iii) ∴ C. S. is $y = C_1 x + x^{-\frac{1}{2}}\left[C_2 \cos\dfrac{\sqrt{3}}{2}(\log x) + C_3 \sin\dfrac{\sqrt{3}}{2}(\log x) \right] + x\log x + 7$

Example 70: Solve $x^2 \dfrac{d^2y}{dx^2} - x\dfrac{dy}{dx} + 4y = \cos(\log x) + x\sin(\log x)$

Solution: Given D. E. is $x^2 \dfrac{d^2y}{dx^2} - x\dfrac{dy}{dx} + 4y = \cos(\log x) + \sin(\log x)$ ... ... (1)

This is Cauchy's D. E., $\therefore$ Put $x = e^z$, $z = \log x$, $\dfrac{d}{dz} = D$

$\therefore x^2 \dfrac{d^2y}{dx^2} = D(D-1)y$; $x\dfrac{dy}{dx} = Dy$

$\therefore$ Equation (1) $\rightarrow [D(D-1) - D + 4]\, y = \cos z + x\sin z$

$\therefore (\mathbf{D^2 - 2D + 4})\mathbf{y} = \cos \mathbf{z} + x\sin \mathbf{z}$ ... Which is L. D. E. with constant coefficient

$\therefore$ C. S. = C. F. + P. I.

i) To find C. F.

Its A. E. is $D^2 - 2D + 4 = 0$

$\therefore D = 1 \pm i\sqrt{3}$

$\therefore$ C. F. $= e^z\left(C_1 \cos\sqrt{3}\,z + C_2 \sin\sqrt{3}\,z\right) = x\left[C_1 \cos\sqrt{3}(\log x) + C_2 \sin\sqrt{3}(\log x)\right]$

ii) To find P. I.

$$P. I. = \frac{1}{D^2 - 2D + 4}(\cos z + e^z \sin z)$$

$$= \frac{1}{D^2 - 2D + 4}\cos z + \frac{1}{D^2 - 2D + 4} e^z \sin z$$

$$= \frac{1}{-1 - 2D + 4}\cos z + e^z \frac{1}{(D+1)^2 - 2(D+1) + 4}\sin z$$

$$= \frac{1}{3 - 2D}\cos z + e^z \frac{1}{D^2 + 2D + 1 - 2D - 2 + 4}\sin z$$

$$= \frac{(3 + 2D)}{9 - 4D^2}\cos z + e^z \frac{1}{D^2 + 3}\sin z$$

$$= \frac{(3 + 2D)\cos z}{9 - 4(-1)} + e^z \frac{\sin z}{-1 + 3}$$

$$= \left[\frac{3\cos z + 2(-\sin z)}{13}\right] + \frac{e^z}{2}\sin z$$

$$= \frac{3}{13}\cos(\log x) - \frac{2}{13}\sin(\log x) + \frac{x}{2}\sin(\log x)$$

iii) $\therefore$ C. S. is $y = x\left[C_1 \cos\sqrt{3}(\log x) + C_2 \sin\sqrt{3}(\log x)\right]$

$$+ \frac{3}{13}\cos(\log x) - \frac{2}{13}\sin(\log x) + \frac{x}{2}\sin(\log x)$$

Example 71: Solve $r\dfrac{d^2y}{dr^2} + \dfrac{dy}{dr} - \dfrac{y}{r} = -ar^2$

Solution: Given D. E. on multiplying bothsides by r isr$^2 \dfrac{d^2y}{dr^2} + r\dfrac{dy}{dr} - y = -ar^3$ ... (1)

This is Cauchy's D. E.,   ∴ Put $r = e^z$, $z = \log r$, $\dfrac{d}{dr} = D$

∴ $r^2 \dfrac{d^2y}{dr^2} = D(D-1)y$; $r\dfrac{dy}{dr} = Dy$

∴ Equation(1)→$[D(D-1) + D - 1]y = -ae^{3z}$

$\left(D^2 - 1\right)y = -ae^{3z}$ ... Which is L. D. E. with constant coefficient

∴ C. S. = C. F. + P. I.

i) **To find C. F.**

Its A. E. is $D^2 - 1 = 0$; $D^2 = 1$; $D = \pm 1 = 1, \ -1$

∴ C. F. = $C_1 e^z + C_2 e^{-z} = C_1 r + C_2 r^{-1}$

ii) **To find P. I.**

P. I. $= \dfrac{1}{D^2 - 1}(-ae^{3z})$

$= \dfrac{-a\,e^{3z}}{(3)^2 - 1} = \dfrac{-a}{8}e^{3z} = \dfrac{-a}{8}r^3$

iii) **C. S. is** $y = C_1 r + C_2 r^{-1} - \dfrac{a}{8}r^3$

Example 72: Solve $\dfrac{d^2v}{dx^2} + \dfrac{1}{r}\dfrac{dv}{dr} = A + B\log r$

Solution: Given D. E. can be written as $r^2\dfrac{d^2v}{dr^2} + r\dfrac{dv}{dr} = Ar^2 + Br^2\log r$ ... ... (1)

This is Cauchy's D. E.,   ∴ Put $r = e^z$, $z = \log r$, $\dfrac{d}{dr} = D$

∴ $r^2\dfrac{d^2v}{dr^2} = D(D-1)y$; $r\dfrac{dv}{dr} = Dy$

∴ Equation: (1)→$[D(D-1) + D]v = Ae^{2z} + Be^{2z}.z$

∴ $D^2v = A\,e^{2z} + B\,ze^{2z}$ ... Which is L. D. E. with constant coefficient.

∴ C. S. = C. F. + P. I.

i) **To find C. F.**

Its A. E. is $D^2 = 0$; $D = 0,0$

∴ C. F. = $C_1 + C_2 z = C_1 + C_2 \log r$

ii) **To find P. I.**

P. I. $= \dfrac{1}{D^2}(A\,e^{2z} + B\,ze^{2z})$

$= A\dfrac{1}{D^2}e^{2z} + B\dfrac{1}{D^2}ze^{2z}$

$$= A\frac{e^{2z}}{4} + B\frac{1}{D}\int ze^{2z}dz$$

$$= \frac{A}{4}e^{2z} + B\frac{1}{D}\left(z\frac{e^{2z}}{2} - \frac{e^{2z}}{4}\right)$$

$$= \frac{A}{4}e^{2z} + B\left[z\frac{e^{2z}}{4} - \frac{e^{2z}}{8} - \frac{e^{2z}}{8}\right]$$

$$= \frac{A}{4}e^{2z} + \frac{B}{4}ze^{2z} - \frac{B}{4}e^{2z}$$

$$= \frac{A}{4}e^{2z} + \frac{B}{4}e^{2z}(z-1)$$

$$= \frac{A}{4}r^2 + \frac{B}{4}r^2(\log r - 1)$$

iii) $\therefore$ C. S. is $\quad y = C_1 + C_2\log r + \dfrac{A}{4}r^2 + \dfrac{B}{4}r^2(\log r - 1)$

Example 73: Solve $x^2\dfrac{d^2y}{dx^2} - 3x\dfrac{dy}{dx} + 5y = x^2\log x$

Solution: Given D. E. is $x^2\dfrac{d^2y}{dx^2} - 3x\dfrac{dy}{dx} + 5y = x^2\log x$ $\qquad$ ... ... (1)

This is Cauchy's D. E., $\qquad$ Put $x = e^z$, $\qquad z = \log x$, $\qquad \dfrac{d}{dz} = D$

$\therefore x^2\dfrac{d^2y}{dx^2} = D(D-1)y$; $\quad x\dfrac{dy}{dx} = Dy$

$\therefore$ Equation(1)$\rightarrow[D(D-1) - 3D + 5]y = e^{2z}z$

$\therefore (D^2 - 4D + 5)y = ze^{2z}$ $\qquad$ ...Which is L. D. E. with constant coefficient

$\therefore$ C. S. = C. F. + P. I.

i) To find C. F.

Its A. E. $D^2 - 4D + 5 = 0$; $\quad D = 2 \pm i$

$\therefore$ C. F. = $e^{2z}(C_1\cos z + C_2\sin z)$

$\therefore$ C. F. = $x^2[C_1\cos(\log x) + C_2\sin(\log x)]$

ii) To find P. I.

$$\text{P. I. } = \frac{1}{D^2 - 4D + 5}\cdot ze^{2z}$$

$$= e^{2z}\frac{1}{(D+2)^2 - 4(D+2) + 5}z$$

$$= e^{2z}\frac{1}{D^2 + 4D + 4 - 4D - 8 + 5}z$$

$$= e^{2z}\frac{1}{D^2 + 1}z = e^{2z}(1 + D^2)^{-1}z$$

$$= e^{2z}(1 - D^2 + \cdots)z$$
$$= e^{2z}(z)$$
$$= (x)^2 \log x$$

iii) $\therefore$ C.S. is $y = x^2[C_1 \cos(\log x) + C_2 \sin(\log x)] + x^2 \log x$

Example 74: Evaluate $4x^2 \dfrac{d^2y}{dx^2} + x \dfrac{dy}{dx} - y = x + \log x$

Solution: Given D.E. is $4x^2 \dfrac{d^2y}{dx^2} + x \dfrac{dy}{dx} - y = x + \log x$ ...... (1)

This is Cauchy's D.E., $\therefore$ Put $x = e^z$, $z = \log x$, $\dfrac{d}{dz} = D$

$\therefore x^2 \dfrac{d^2y}{dx^2} = D(D-1)y$; $x \dfrac{dy}{dx} = Dy$

$\therefore$ Equation(1)$\rightarrow[4D(D-1) + D - 1]y = e^z + z$

$(4D^2 - 4D + D - 1)y = e^z + z$

$(4D^2 - 3D - 1)y = e^z + z$ ... Which is L.D.E. with constant coefficient

$\therefore$ C.F. = C.F. + P.I.

i) To find C.F.

Its A.E. is $4D^2 - 3D - 1 = 0$; $D = 1, \dfrac{-1}{4}$

$\therefore$ C.F. = $C_1 e^z + C_2 e^{-\frac{1}{4}z} = C_1 x + C_2 x^{-\frac{1}{4}}$

ii) To find P.I.

$$P.I. = \dfrac{1}{4D^2 - 3D - 1}(e^z + z)$$

$$= \dfrac{1}{4D^2 - 3D - 1} e^z + \dfrac{1}{4D^2 - 3D - 1} z$$

$$= \dfrac{z}{8D - 3} e^z + \dfrac{1}{-[1 - (4D^2 - 3D)]} z$$

$$= \dfrac{ze^z}{8(1) - 3} - [1 - (4D^2 - 3D)]^{-1} z$$

$$= \dfrac{ze^z}{5} - [1 + (4D^2 - 3D) + \cdots] z$$

$$= \dfrac{ze^z}{5} - z + 3$$

$$= x \dfrac{\log x}{5} - \log x + 3$$

iii) $\therefore$ C.S. is $y = C_1 x + C_2 x^{-\frac{1}{4}} + \dfrac{x \log x}{5} - \log x + 3$

Example 75: Solve $(x^3D^3 + 3x^2D^2 + xD)y = 24x^2$

Solution: Given D. E. is $(x^3D^3 + 3x^2D^2 + xD)y = 24x^2$ ... ... (1)

This is cauchy's D. E.,   ∴ Put $x = e^z$, $z = \log x$, $\dfrac{d}{dz} = \theta$ ... ... (2)

∴ $x^3D^3 = \theta(\theta - 1)(\theta - 2)y$; $x^2D^2 = \theta(\theta - 1)y$; $xD = \theta y$

∴ $[\theta^3 - \theta^2 - 2\theta^2 + 2\theta + 3\theta^2 - 3\theta + \theta]y = 24e^{2z}$

$\theta^3 = 24e^{2z}$ ... Which is L. D. E. with constant coefficient

∴ C. S. = C. F. + P. I.

i) **To find C. F.**

Its A. E. is $\theta^3 = 0$; $\theta = 0, 0, 0$.

∴ C. F. = $C_1 + C_2z + C_3z^2 = C_1 + C_2\log x + C_3(\log x)^2$

ii) **To find P. I.**

P. I. $= \dfrac{1}{\theta^3}24e^{2z}$

$= 24\dfrac{1}{\theta^3}e^{2z}$ $= 24\dfrac{e^{2z}}{2^3}$ $= 3e^{2z}$ $= 3x^2$

iii) ∴ **C. S. is** $y = C_1 + C_2\log x + C_3(\log x)^2 + 3x^2$

Example 76: Solve $\left(x^2\dfrac{d^2y}{dx^2} - 3x\dfrac{dy}{dx} + y\right) = \dfrac{1}{x}\sin(\log x)$

Solution: Given D. E. is $x^2\dfrac{d^2y}{dx^2} - 3x\dfrac{dy}{dx} + y = \dfrac{1}{x}\sin(\log x)$ ... ... (1)

This is cauchy's D. E.,   ∴ Put $x = e^z$, $z = \log x$, $\dfrac{d}{dz} = D$

∴ $x^2\dfrac{d^2y}{dx^2} = D(D - 1)y$; $x\dfrac{dy}{dx} = Dy$

∴ Equation(1)→$[D(D - 1) - 3D + 1]y = e^{-z}\sin(z)$

$(D^2 - 4D + 1)y = e^{-z}\sin(z)$ ... Which is L. D. E. with constant coefficient

∴ C. S. = C. F. + P. I.

i) **To find C. F.**

Its A. E. is $D^2 - 4D + 1 = 0$; $D = 2 \pm \sqrt{3} = 2 + \sqrt{3},\ 2 - \sqrt{3}$

∴ C. F. = $C_1e^{(2+\sqrt{3})z} + C_2e^{(2-\sqrt{3})z} = e^{2z}\left(C_1e^{\sqrt{3}z} + C_2e^{-\sqrt{3}z}\right)$

∴ C. F. = $x^2\left(C_1x^{\sqrt{3}} + C_2x^{-\sqrt{3}}\right)$

ii) **To find P. I.**

P. I. $= \dfrac{1}{D^2 - 4D + 1}e^{-z}\sin z$

$= e^{-z}\dfrac{1}{(D-1)^2 - 4(D-1) + 1}\sin z$

$$= e^{-z} \frac{1}{D^2 - 2D + 1 - 4D + 4 + 1} \sin z$$

$$= e^{-z} \frac{1}{D^2 - 6D + 6} \sin z$$

$$= e^{-z} \frac{1}{-1 - 6D + 6} \sin z$$

$$= e^{-z} \frac{1}{5 - 6D} \sin z$$

$$= e^{-z} \frac{(5 + 6D)}{25 - 36D^2} \sin z$$

$$= e^{-z} \frac{(5 + 6D).\sin z}{25 - 36(-1)} \quad = \frac{e^{-z}}{61}(5 \sin z + 6D \sin z) \quad = \frac{e^{-z}}{61}(5 \sin z + 6 \cos z)$$

$$= \frac{x^{-1}}{61}(5 \sin(\log x) + 6 \cos(\log x))$$

iii) $\therefore$ **C. S. is** $\quad y = x^2(C_1 x^{\sqrt{3}} + C_2 x^{-\sqrt{3}}) + \dfrac{x^{-1}}{61}(5 \sin \log x + 6 \cos \log x)$

Example 77: Evaluate $x \dfrac{d^3 y}{dx^3} + 2 \dfrac{d^2 y}{dx^2} = x^2$

Solution: Given D. E. can be written as $x^3 \dfrac{d^3 y}{dx^3} + 2x^2 \dfrac{d^2 y}{dx^2} = x^4$ ... ... (1)

This is Cauchys D. E., $\therefore$ Put $x = e^z$, $z = \log x$, $\dfrac{d}{dz} = D$

$\therefore x^3 \dfrac{d^3 y}{dx^3} = D(D - 1)(D - 2)y$; $x^2 \dfrac{d^2 y}{dx^2} = D(D - 1)y$

$\therefore$ Equation(1)$\rightarrow [D(D - 1)(D - 2) + 2D(D - 1)]y = e^{4z}$

$[D^3 - 2D^2 - D^2 + 2D + 2D^2 - 2D]y = e^{4z}$

$(D^3 - D^2) y = e^{4z}$ ... Which is L. D. E. with constant coefficient

   C. S. = C. F. + P. I.

i) To find C. F

Its A. E. is $D^3 - D^2 = 0$; $D^2(D - 1) = 0$; $D = 0, 0, 1$

$\therefore$ C. F. $= C_1 + C_2 z + C_3 e^z = C_1 + C_2 \log x + C_3 x$

ii) To find P. I.

$\therefore$ P. I. $= \dfrac{1}{D^3 - D^2} e^{4z}$

$$= \frac{e^{4z}}{4^3 - 4^2} \quad = \frac{e^{4z}}{64 - 16} \quad = \frac{e^{4z}}{48} \quad = \frac{x^4}{48}$$

iii) $\therefore$ **C. S. is** $\quad y = C_1 + C_2 \log x + C_3 x + \dfrac{x^4}{48}$

Example 78: Solve $x^2 \dfrac{d^2y}{dx^2} = 2y + \dfrac{1}{x}$

Solution: Given D. E. can be written as $x^2 \dfrac{d^2y}{dx^2} - 2y = \dfrac{1}{x}$ ... ... (1)

This is Cauchy's D. E.,   ∴ Put $x = e^z$, $z = \log x$, $\dfrac{d}{dz} = D$

∴ $x^2 \dfrac{d^2y}{dx^2} = D(D-1)y$.

∴ Equation (1) → $[D(D-1) - 2]y = \dfrac{1}{e^z}$

$(D^2 - D - 2)y = e^{-z}$ ... Which is L. D. E. with constant coefficient

∴ C. S. = C. F. + P. I.

i) **To find C. P.**

Its A. E. is $D^2 - D - 2 = 0$; $D = -1,\ 2$

∴ C. F. = $C_1 e^{-z} + C_2 e^{2z} = C_1 x^{-1} + C_2 x^2$

ii) **To find P. I.**

P. I. $= \dfrac{1}{D^2 - D - 2} e^{-z}$

$= \dfrac{z}{2D - 1} e^{-z}$

$= \dfrac{z\, e^{-z}}{2(-1) - 1} = \dfrac{z\, e^{-z}}{-3} = \dfrac{(\log x)\, x^{-1}}{-3} = -\dfrac{(\log x)}{3x}$

iii) ∴ C. S. is, $y = C_1 x^{-1} + C_2 x^2 - \dfrac{(\log x)}{3x}$

Example 79: Solve $\left(\dfrac{d^2}{dx^2} - \dfrac{2}{x^2}\right)^2 y = 0$

Solution: Given D. E. $\left(\dfrac{d^2}{dx^2} - \dfrac{2}{x^2}\right)\left(\dfrac{d^2}{dx^2} - \dfrac{2}{x^2}\right) y = 0$

$\left(\dfrac{d^2}{dx^2} - \dfrac{2}{x^2}\right) v = 0$ ... ... (1)

Consider $\left(\dfrac{d^2}{dx^2} - \dfrac{2}{x^2}\right) y = v$ ... ... (2)

∴ Equation(1) → $\left(\dfrac{d^2}{dx^2} - \dfrac{2}{x^2}\right) v = 0$

i. e. $\dfrac{d^2v}{dx^2} - \dfrac{2v}{x^2} = 0$

i.e. $x^2 \dfrac{d^2v}{dx^2} - 2v = 0$ {$\because$ Multiplying by x^2 on both sides   ... ... (3)

This is Cauchy's D. E.

$\therefore$ Put $x = e^z$, $\quad$ $z = \log x$, $\quad$ $\dfrac{d}{dz} = D$

$\therefore x^2 \dfrac{d^2v}{dx^2} = D(D-1)v$

$\therefore$ Equation (3) $\rightarrow [D(D-1) - 2]v = 0$

$(D^2 - D - 2)v = 0$ ... Which is L. D. E. with constant coefficient

$\therefore$ $\quad$ C. S. $= v = $ C. F. $+$ P. I.

To find C. F.

Its A. E. is $D^2 - D - 2 = 0$; $\quad$ $D = -1,\ 2$

$\therefore$ C. F. $=$ C. S. $= v = C_1 e^{-z} + C_2 e^{2z}$

$\qquad\qquad\quad v = C_1 x^{-1} + C_2 x^2$

$\quad$ P. I. $= 0$

$\therefore$ Equation$(2) \rightarrow \left(\dfrac{d^2}{dx^2} - \dfrac{2}{x^2} \right) y = C_1 x^{-1} + C_2 x^2$

$\therefore \qquad \dfrac{d^2y}{dx^2} - \dfrac{2y}{x^2} = C_1 x^{-1} + C_2 x^2$

i. e. $x^2 \dfrac{d^2y}{dx^2} - 2y = C_1 x + C_2 x^4$ ... ... (4)

This is Cauchy's D. E.

$\therefore$ Put $x = e^t$, $\quad$ $t = \log x$, $\quad$ $\dfrac{d}{dt} = \theta$

$\therefore x^2 \dfrac{d^2y}{dx^2} = \theta(\theta - 1)y$

$\therefore$ Equation$(4) \rightarrow [\theta(\theta-1) - 2]y = C_1 e^t + C_2 e^{4t}$

$(\theta^2 - \theta - 2)y = C_1 e^t + C_2 e^{4t}$ ... Which is L. D. E. with constant coefficient

i) To find C. F.

Its A. E. $\theta^2 - \theta - 2 = 0$; $\quad$ $\theta = -1,\ 2$

$\therefore$ C. F. $= C_3 e^{-t} + C_4 e^{2t} = C_3 x^{-1} + C_4 x^2$

ii) To find P. I.

P. I. $= \dfrac{1}{\theta^2 - \theta - 2}[C_1 e^t + C_2 e^{4t}]$

$\qquad = \dfrac{1}{\theta^2 - \theta - 2} C_1 e^t + \dfrac{1}{\theta^2 - \theta - 2} C_2 e^{4t}$

$\qquad = C_1 \dfrac{e^t}{1 - 1 - 2} + C_2 \dfrac{e^{4t}}{4^2 - 4 - 2}$

$$= C_1 \frac{e^t}{-2} + C_2 \frac{e^{4t}}{10}$$

$$= \frac{-1}{2} C_1 x + \frac{1}{10} C_2 x^4$$

iii) $\therefore$ C.S. is $y = \frac{C_3}{x} + C_4 x^2 - \frac{1}{2} C_1 x + \frac{1}{10} C_2 x^4$

Example 80: Solve $x^2 \dfrac{d^2y}{dx^2} - x \dfrac{dy}{dx} + y = x \log x$

Soiution: Given D. E. is $\quad x^2 \dfrac{d^2y}{dx^2} - x \dfrac{dy}{dx} + y = x \log x \quad$ ... ... (1)

This is Cauchy's D. E., $\qquad \therefore$ Put $x = e^z$, $\quad z = \log x$, $\quad \dfrac{d}{dz} = D$

$\therefore x^2 \dfrac{d^2y}{dx^2} = D(D-1)y \;\; ; \quad x \dfrac{dy}{dx} = Dy$

$\therefore$ Equation(1)$\rightarrow [D(D-1) - D + 1]y = e^z z$

$\therefore (D^2 - 2D + 1)y = ze^z$

   i.e. $(D-1)^2 y = ze^z \qquad$ ... Which is L. D. E. with constant coefficient

$\therefore$ C. S. = C. F. + P. I.

i) To find C. F.

Its A. E. is $(D-1)^2 = 0 \; ; \; D = 1,1.$

$\therefore$ C. F. = $(C_1 + C_2 z)e^z = (C_1 + C_2 \log x) x$

ii) To find P. I.

$\text{P. I.} = \dfrac{1}{(D-1)^2} ze^z$

$\qquad = e^z \dfrac{1}{[(D+1) - 1]^2} z$

$\qquad = e^z \dfrac{1}{D^2} z \;\; = e^z \dfrac{z^3}{6} \;\; = x \dfrac{(\log x)^3}{6}$

iii) $\therefore$ **C. S. is** $\quad y = (C_1 + C_2 \log x) x + \dfrac{x}{6}(\log x)^3$

Example 81: Solve $x^2 \dfrac{d^2y}{dx^2} - 2x \dfrac{dy}{dx} - 4y = x^2 + 2 \log x$

Solution: Given D. E. is $\quad x^2 \dfrac{d^2y}{dx^2} - 2x \dfrac{dy}{dx} - 4y = x^2 + 2 \log x \quad$ ... ... (1)

This is Cauchy's D. E., $\qquad \therefore$ Put $x = e^z$, $\quad z = \log x$, $\quad \dfrac{d}{dz} = D$

$\therefore x^2 \dfrac{d^2y}{dx^2} = D(D-1)y \;\; ; \quad x \dfrac{dy}{dx} = Dy$

$\therefore$ Equation $(1) \rightarrow [D(D-1) - 2D - 4]y = e^{2z} + 2z$

$(D^2 - 3D - 4)y = e^{2z} + 2z$ ... Which is L. D. E. with constant coefficient

$\therefore$ C. S. = C. F. + P. I.

i) To find C. F.

Its A. E. is $D^2 - 3D - 4 = 0$; $D = -1, 4$

$\therefore$ C. F. $= C_1 e^{-z} + C_2 e^{4z} = C_1 x^{-1} + C_2 x^4$

ii) To find P. I.

$$\text{P. I.} = \frac{1}{D^2 - 3D - 4}(e^{2z} + 2z)$$

$$= \frac{1}{D^2 - 3D - 4}e^{2z} + 2\frac{1}{D^2 - 3D - 4}z$$

$$= \frac{e^{2z}}{2^2 - 3(2) - 4} + 2\frac{1}{-4\left(1 - \frac{D^2 - 3D}{4}\right)}z$$

$$= \frac{e^{2z}}{-6} - \frac{1}{2}\left(1 - \frac{D^2 - 3D}{4}\right)^{-1}z$$

$$= \frac{e^{2z}}{-6} - \frac{1}{2}\left(1 + \frac{D^2 - 3D}{4} + \cdots\right)z$$

$$= \frac{-1}{6}e^{2z} - \frac{1}{2}\left(z - \frac{3}{4}\right)$$

$$= \frac{-1}{6}x^2 - \frac{1}{2}\left(\log x - \frac{3}{4}\right) \qquad = \frac{-1}{6}x^2 - \frac{\log x}{2} + \frac{3}{8}$$

iii) $\therefore$ C. S. is $y = C_1 x^{-1} + C_2 x^4 - \frac{1}{6}x^2 - \frac{\log x}{2} + \frac{3}{8}$

7. ii   Legendre's linear equation

An equation of the form

$$k_0(ax+b)^n\frac{d^n y}{dx^n} + k_1(ax+b)^{n-1}\frac{d^{n-1}y}{dx^{n-1}} + \cdots + k_n y = X$$

Where $k's, a, b$ are constants &X is a function of x, is called Legendre's linear equation.

Such equation can be reduced to linear equations with constant coefficient by the substitution

$$ax + b = e^z, \qquad \text{i. e.}\quad z = \log(ax+b), \qquad \frac{d}{dz} = D$$

$$(ax+b)^3\frac{d^3 y}{dx^3} = a^3 D(D-1)(D-2)y$$

$$(ax + b)^2 \frac{d^2y}{dx^2} = a^2 D(D - 1)y$$

$$(ax + b)\frac{dy}{dx} = aDy \qquad \text{... and so on}$$

After making these replacements in equation (1) there result a linear equation with constant coefficients.

7. ii. a   Examples onLegendre's linear equation

Example 82: Solve $(1 + x)^2 \frac{d^2y}{dx^2} + (1 + x)\frac{dy}{dx} + y = 2\sin[\log(1 + x)]$

Solution: Given D. E. $(1 + x)^2 \frac{d^2y}{dx^2} + (1 + x)\frac{dy}{dx} + y = 2\sin[\log(1 + x)]$ ... ... (1)

This is Legendre's D. E.

$\therefore$ Put $(1 + x) = e^z$; $z = \log(1 + x)$; $\dfrac{d}{dz} = D$

$\therefore (1 + x)^2 \dfrac{d^2y}{dx^2} = D(D - 1)\,y$; $(1 + x)\dfrac{dy}{dx} = Dy$

Equation(1)$\rightarrow [\,D(D - 1) + D + 1]y = 2\sin(z)$

$(D^2 + 1)y = 2\sin(z) \qquad$ ... which is L. D. E. with constant coefficient

$\therefore$ C. S. = C. F. + P. I.

i) To find C. F.

Its A. E. is $D^2 + 1 = 0$; $D^2 = -1$; $D = \pm i$

$\therefore$ C. F. = $C_1 \cos z + C_2 \sin z = C_1 \cos\big(\log(x + 1)\big) + C_2 \sin\big(\log(x + 1)\big)$

ii) To find P. I.

$$\text{P. I.} = \frac{1}{D^2 + 1}\,2\sin z = 2\frac{z}{2D}\sin z = z\int \sin z \; dz$$

$$= -z\cos z$$

$$= -\log(1 + x)\cos[\log(1 + x)]$$

iii) $\therefore$ **C. S. is**

$$y = C_1 \cos[\log(1 + x)] + C_2 \sin[\log(1 + x)] - \log(1 + x)\cos[(1 + x)]$$

Example 83: Solve $(2x - 1)^2 \frac{d^2y}{dx^2} + (2x - 1)\frac{dy}{dx} - 2y = 8x^2 - 2x + 3$

Solution: Given D. E. $(2x - 1)^2 \frac{d^2y}{dx^2} + (2x - 1)\frac{dy}{dx} - 2y = 8x^2 - 2x + 3$ ... ... (1)

This is Legendre's D. E.

$\therefore$ Put $(2x - 1) = e^z$; $z = \log(2x - 1)$; $x = \dfrac{e^z + 1}{2}$; $\dfrac{d}{dz} = D$

$$\therefore (2x-1)^2 \frac{d^2y}{dx^2} = 4D(D-1)y; \quad (2x-1)\frac{dy}{dx} = 2Dy$$

$$\therefore \text{Equation}(1) \rightarrow [4D(D-1) + 2D - 2]y = 8\left(\frac{e^z+1}{2}\right)^2 - 2\left(\frac{e^z+1}{2}\right) + 3$$

$$\therefore (4D^2 - 2D - 2)y = \frac{8}{4}(e^z+1)^2 - (e^z+1) + 3$$

$$2(2D^2 - D - 1)y = 2(e^{2z} + 2e^z + 1) - e^z - 1 + 3 = 2e^{2z} + 3e^{3z} + 4$$

$$(2D^2 - D - 1)y = e^{2z} + \frac{3e^z}{2} + 2 \text{ ... Which is L. D. E. with constant coefficient}$$

$$\therefore \text{C. S.} = \text{C. F.} + \text{P. I.}$$

i) To find C. F.

Its A. E. is $2D^2 - D - 1 = 0$; $D = 1, \dfrac{-1}{2}$

$$\therefore \text{C. F.} = C_1 e^z + C_2 e^{-\frac{z}{2}}$$

$$\text{C. F.} = C_1(2x-1) + C_2(2x-1)^{-\frac{1}{2}}$$

ii) To find P. I.

$$\text{P. I.} = \frac{1}{2D^2 - D - 1}\left(e^{2z} + \frac{3}{2}e^z + 2\right)$$

$$= \frac{1}{2D^2 - D - 1}e^{2z} + \frac{3}{2}\frac{1}{2D^2 - D - 1}e^z + 2\frac{1}{2D^2 - D - 1}e^{-z}$$

$$= \frac{e^{2z}}{2(2)^2 - 2 - 1} + \frac{3}{2}\frac{z}{4D - 1}e^z + 2\frac{e^{0z}}{2(0)^2 - 0 - 1}$$

$$= \frac{e^{2z}}{5} + \frac{3}{2}\frac{ze^z}{4(1) - 1} + \frac{2}{-1}$$

$$= \frac{e^{2z}}{5} + \frac{ze^z}{2} - 2$$

$$= \frac{(2x-1)^2}{5} + \frac{1}{2}\log(2x-1)\,[(2x-1)] - 2$$

iii) ∴ C. S. is

$$y = C_1(2x-1) + C_2(2x-1)^{-\frac{1}{2}} + \frac{1}{5}(2x-1)^2 + \frac{1}{2}(2x-1)\log(2x-1) - 2$$

Example 84: Evaluate $(2x+1)^2\dfrac{d^2y}{dx^2} - 2(2x+1)\dfrac{dy}{dx} - 12y = 6x$

Solution: Given D. E. $\quad (2x+1)^2\dfrac{d^2y}{dx^2} - 2(2x+1)\dfrac{dy}{dx} - 12y = 6x \qquad \text{... ... (1)}$

This is Legendre's D. E.

$\therefore$ Put $(2x + 1) = e^z$; $z = \log(2x + 1)$; $x = \dfrac{e^z - 1}{2}$; $\dfrac{d}{dz} = D$

$\therefore (2x + 1)^2 \dfrac{d^2y}{dx^2} = 4D(D - 1)y$; $(2x + 1)\dfrac{dy}{dx} = 2Dy$

$\therefore$ Equation(1)$\rightarrow [4D(D - 1) - 2(2D) - 12]y = 6\left(\dfrac{e^z - 1}{2}\right)$

$\therefore (4D^2 - 4D - 4D - 12)y = 3(e^z - 1)$

$(4D^2 - 8D - 12)y = 3(e^z - 1)$

$(D^2 - 2D - 3)\, y = \dfrac{3}{4}(e^z - 1)$ ... Which is L. D. E. with constant coefficient

$\therefore$ C. S. = C. F. + P. I.

i) To find C. F.

Its A. E. is $(D^2 - 2D - 3) = 0$; $D = 3, -1$

$\therefore$ C. F. = $C_1 e^{3z} + C_2 e^{-z} = C_1(2x + 1)^3 + C_2(2x + 1)^{-1}$

ii) To find P. I.

$\therefore$ P. I. = $\dfrac{1}{D^2 - 2D - 3}\dfrac{3}{4}(e^z - 1)$

$= \dfrac{3}{4}\left[\dfrac{1}{D^2 - 2D - 3}e^z - \dfrac{1}{D^2 - 2D - 3}e^{0z}\right]$

$= \dfrac{3}{4}\left[\dfrac{e^z}{(1)^2 - 2(1) - 3} - \dfrac{1}{0^2 - 2(0) - 3}\right]$

$= \dfrac{3}{4}\left[\dfrac{e^z}{-4} - \dfrac{1}{-3}\right]$ $= \dfrac{-3e^z}{16} + \dfrac{1}{4}$ $= \dfrac{-3\,(2x + 1)}{16} + \dfrac{1}{4}$ $= \dfrac{-3}{8}x - \dfrac{3}{16} + \dfrac{1}{4}$

$= -\dfrac{3}{8}x + \dfrac{1}{16}$

iii) $\therefore$ **C. S. is** $y = C_1(2x + 1)^3 + C_2(2x + 1)^{-1} - \dfrac{3}{8}x + \dfrac{1}{16}$

Example 85: Solve $(3x + 2)^2 \dfrac{d^2y}{dx^2} + 3(3x + 2)\dfrac{dy}{dx} + 36y = 3x^2 + 4x + 1$

Solution: Given D. E. is $(3x + 2)^2 \dfrac{d^2y}{dx^2} + 3(3x + 2)\dfrac{dy}{dx} + 36y = 3x^2 + 4x + 1$

This is Legenders D. E.

Put $3x + 2 = e^z$, $x = \dfrac{e^z - 2}{3}$, $z = \log(3x + 2)$, $\dfrac{d}{dz} = D$

$\therefore (3x + 1)^2 \dfrac{d^2y}{dx^2} = 9D(D - 1)y$; $(3x + 2)\dfrac{dy}{dx} = 3dy$

$\therefore$ Equation (1)$\rightarrow [9D(D - 1) + 3(3D) + 36]y = 3\left(\dfrac{e^z - 2}{3}\right)^2 + 4\left(\dfrac{e^z - 2}{3}\right) + 1$

$$(9D^2 + 36)y = \frac{1}{3}(e^{2z} - 4e^z + 4) + \frac{4}{3}(e^z - 2) + 1$$

$$9(D^2 + 4)y = \frac{1}{3}[e^{2z} - 4e^z + 4 + 4e^z - 8 + 3]$$

$$(D^2 + 4)y = \frac{1}{27}[e^{2z} - 1] \qquad \text{... Which is LDE with constant coefficient}$$

∴ C. S. = C. F. + P. I.

i) To find C. F.

Its A. E. is $D^2 + 4 = 0$; $D^2 = -4$; $D = \pm 2i$

∴ C. F. = $C_1 \cos 2z + C_2 \sin 2z = C_1 \cos 2[\log(3x + 2)] + C_2 \sin 2[\log(3x + 2)]$

ii) To find P. I.

$$\therefore \quad \text{P. I.} = \frac{1}{D^2 + 4} \frac{1}{27}(e^{2z} - 1)$$

$$= \frac{1}{27}\left[\frac{1}{D^2 + 4}e^{2z} - \frac{1}{D^2 + 4}e^{0z}\right]$$

$$= \frac{1}{27}\left[\frac{e^{2z}}{2^2 + 4} - \frac{e^{0z}}{0^2 + 4}\right] = \frac{1}{27}\left[\frac{e^{2z}}{8} - \frac{1}{4}\right] = \frac{1}{108}\left(\frac{1}{2}e^{2z} - 1\right)$$

$$= \frac{1}{108}\left[\frac{1}{2}(3x + 2)^2 - 1\right]$$

iii) ∴ C. S. is

$$y = C_1 \cos 2[\log(3x + 2)] + C_2 \sin 2[\log(3x + 2)] + \frac{1}{108}\left[\frac{1}{2}(3x + 2)^2 - 1\right]$$

8   TWO OTHER METHODS OF FINDING P. I.

    i.   **Method of variation of paramenters (Wronskian Method)**

    ii.   **Method of undetermind coefficient**

8.i Method of Variation of Parameters [VOP]

Steps to solve this type of problems:

For 2nd order D. E.	For 3rd order D. E.

For 2nd order D. E.

i) Compare its C. F. with

$C.F. = C_1 y_1 + C_2 y_2$

where $y_1 \& y_2$ are function of x

ii) To find W

$$W = \begin{vmatrix} y_1 & y_2 \\ y_1' & y_2' \end{vmatrix}$$

iii) Its P. I. $= U y_1 + V y_2$

Where,

$$U = -\int \frac{y_2}{W} X \, dx \,,$$

$$V = \int \frac{y_1}{W} X \, dx$$

For 3rd order D. E.

i) Compare its C. F. with

$C.F. = C_1 y_1 + C_2 y_2 + C_3 y_3$

where $y_1, y_2 \& y_3$ are function of x

ii) To find W

$$W = \begin{vmatrix} y_1 & y_2 & y_3 \\ y_1' & y_2' & y_3' \\ y_1'' & y_2'' & y_3'' \end{vmatrix}$$

iii) Its P. I. $= U y_1 + V y_2 + M y_3$

Where, $U = \int \dfrac{\begin{vmatrix} y_2 & y_3 \\ y_2' & y_3' \end{vmatrix}}{W}.X \, dx \,,$

$$V = -\int \frac{\begin{vmatrix} y_1 & y_3 \\ y_1' & y_3' \end{vmatrix}}{W}.X \, dx \,,$$

$$M = \int \frac{\begin{vmatrix} y_1 & y_2 \\ y_1' & y_2' \end{vmatrix}}{W}.X \, dx$$

8.i.a Examples on Method of Variation of Parameters [VOP]

Example 86: Using the method of variation of parmeters, solve

$$\frac{d^2 y}{dx^2} + 4y = \tan 2x$$

Solution: Given D. E. in symbolic form $(D^2 + 4) y = \tan 2x$

$\therefore$ C. S. $=$ C. F. $+$ P. I.

i) To find C. F.

Its A. E. is $D^2 + 4 = 0$; $D^2 = -4$; $D = \pm 2i$

$\therefore$ C. F. $= C_1 \cos 2x + C_2 \sin 2x$

ii) To find P. I.

P. I. $= y_1 U + y_2 V$

Where, $U = -\int \dfrac{y_2}{W} X \, dx \,,$ $V = \int \dfrac{y_1}{W} X \, dx \,,$ $W = \begin{vmatrix} y_1 & y_2 \\ y_1' & y_2' \end{vmatrix}$

Now, Compare C. F. $= C_1 y_1 + C_2 y_2$

Here, $y_1 = \cos 2x$; $y_2 = \sin 2x$

$y_1' = -2 \sin 2x$; $y_2' = 2 \cos 2x$

$$W = \begin{vmatrix} y_1 & y_2 \\ y_1' & y_2' \end{vmatrix} = \begin{vmatrix} \cos 2x & \sin 2x \\ -2\sin 2x & 2\cos 2x \end{vmatrix}$$

$$= 2\cos^2 2x + 2\sin^2 2x = 2(\cos^2 2x + \sin^2 2x) \ ; \quad W = 2$$

$$\therefore \quad P.I. = -y_1 \int \frac{y_2}{W} X\,dx + y_2 \int \frac{y_1}{W} X\,dx$$

$$= -y_1 \int \frac{y_2}{W} X\,dx + y_2 \int \frac{y_1}{W} X\,dx$$

$$= -\cos 2x \int \frac{\sin 2x}{2} \tan 2x \,dx + \sin 2x \int \frac{\cos 2x}{2} \tan 2x \,dx$$

$$= -\frac{\cos 2x}{2} \int \frac{\sin^2 2x}{\cos 2x} \,dx + \frac{\sin 2x}{2} \int \sin 2x \,dx$$

$$= -\frac{\cos 2x}{2} \int \frac{1 - \cos^2 2x}{\cos 2x} \,dx + \frac{\sin 2x}{2}\left(\frac{-\cos 2x}{2} \right)$$

$$= -\frac{\cos 2x}{2}\left[\int (\sec 2x - \cos 2x)\,dx + \frac{\sin 2x}{2} \right]$$

$$= -\frac{\cos 2x}{2}\left[\frac{\log(\sec 2x + \tan 2x)}{2} - \frac{\sin 2x}{2} + \frac{\sin 2x}{2} \right]$$

$$= -\frac{1}{4}\cos 2x \, \log(\sec 2x + \tan 2x)$$

iii) ∴ **C.S.** is $y = C_1 \cos 2x + C_2 \sin 2x - \dfrac{1}{4}\cos 2x \log(\sec 2x + \tan 2x)$

Example 87: Solve by VOP, $\dfrac{d^2 y}{dx^2} - y = \dfrac{2}{1 + e^x}$

Solution: Given D.E. in symbolic form is $(D^2 - 1)y = 2(1 + e^x)^{-1}$

C.S. = C.F. + P.I.

i) To find C.F.

Its A.E. is $D^2 - 1 = 0$; $D^2 = 1$; $D = \pm 1 = -1,\ 1$

∴ C.F. $= C_1 e^{-x} + C_2 e^x$

ii) To find P.I.

P.I. $= y_1 U + y_2 V$

Where, $U = -\displaystyle\int \frac{y_2}{W} X\,dx$, $V = \displaystyle\int \frac{y_1}{W} X\,dx$, $W = \begin{vmatrix} y_1 & y_2 \\ y_1' & y_2' \end{vmatrix}$

Now, Compare C.F. $= C_1 y_1 + C_2 y_2$

Here, $y_1 = e^{-x}$; $y_2 = e^x$

$y_1' = -e^{-x}$; $y_2' = e^x$

$$W = \begin{vmatrix} y_1 & y_2 \\ y_1' & y_2' \end{vmatrix} = \begin{vmatrix} e^{-x} & e^x \\ -e^{-x} & e^x \end{vmatrix} = 1 + 1 = 2 \ ; \quad W = 2$$

$$\therefore \quad P.I. = -y_1 \int \frac{y_2}{W} X\,dx + y_2 \int \frac{y_1}{W} X\,dx$$

$$= -e^{-x} \int \frac{e^x}{2} \frac{2}{(1+e^x)} dx + e^x \int \frac{e^{-x}}{2} \frac{2}{(1+e^x)} dx$$

$$= -e^{-x} \int \frac{e^x}{1+e^x} dx + e^x \int \frac{e^{-x}}{1+e^x} dx$$

$$= -e^{-x} \log(1+e^x) + e^x \int \frac{1}{e^x(1+e^x)} dx$$

$$= -e^{-x} \log(1+e^x) + e^x \int \left(\frac{1}{e^x} - \frac{1}{1+e^x} \right) dx \quad \ldots \text{Note}$$

$$= -e^{-x} \log(1+e^x) + e^x \left[\int e^{-x} dx - \int \frac{e^{-x}}{e^{-x}+1} dx \right]$$

$$= -e^{-x} \log(1+e^x) + e^x \left[\frac{e^{-x}}{-1} + \log(e^{-x}+1) \right]$$

$$= -e^{-x} \log(1+e^x) - 1 + e^x \log(e^{-x}+1)$$

iii) $\therefore$ **C.S. is** $\quad y = C_1 e^{-x} + C_2 e^x - 1 - e^{-x} \log(1+e^x) + e^x \log(e^{-x}+1)$

Example 88: Solve by method of VOP $y'' - 6y' + 9y = \dfrac{e^{3x}}{x^2}$

Solution: Given D. E. $y'' - 6y' + 9y = \dfrac{e^{3x}}{x^2}$; Symbolic form $(D^2 - 6D + 9)y = \dfrac{e^{3x}}{x^2}$

C. S. = C. F. + P. I.

i) To find C. F.

Its A. E. is $(D^2 - 6D + 9) = 0$; $(D-3)^2 = 0$; $D = 3,\ 3$

$\therefore$ C. F. = $(C_1 + C_2 x)e^{3x} = C_1 e^{3x} + C_2 x e^{3x}$

ii) To find P. I.

P. I. = $y_1 U + y_2 V$

Where, $U = -\int \dfrac{y_2}{W} X\,dx$, $V = \int \dfrac{y_1}{W} X\,dx$, $W = \begin{vmatrix} y_1 & y_2 \\ y_1' & y_2' \end{vmatrix}$

Now,   Compare   C. F. = $C_1 y_1 + C_2 y_2$

Here, $y_1 = e^{3x}$; $y_2 = x e^{3x}$

$y_1' = 3e^{3x}$; $y_2' = 3x e^{3x} + e^{3x}$

$W = \begin{vmatrix} y_1 & y_2 \\ y_1' & y_2' \end{vmatrix} = \begin{vmatrix} e^{3x} & x e^{3x} \\ 3e^{3x} & 3x e^{3x} + e^{3x} \end{vmatrix} = 3x e^{6x} + e^{6x} - 3x e^{6x}$

$\therefore W = e^{6x}$

$\therefore \quad P. I. = -y_1 \int \dfrac{y_2}{W} X\,dx + y_2 \int \dfrac{y_1}{W} X\,dx$

$$= -e^{3x} \int \frac{xe^{3x} e^{3x}}{e^{6x} x^2} dx + xe^{3x} \int \frac{e^{3x} e^{3x}}{e^{6x} x^2} dx$$

$$= -e^{3x} \int \frac{1}{x} dx + xe^{3x} \int \frac{1}{x^2} dx$$

$$= -e^{3x} \log x + x e^{3x} \left(-\frac{1}{x}\right) = -e^{3x}(\log x + 1)$$

iii) $\therefore$ **C.S. is** $y = (C_1 + C_2 x)e^{3x} - e^{3x}(\log x + 1)$

Example 89: **Solve by method of VOP** $y'' - 2y' + y = e^x \log x$

Solution: Given D. E. in symbolic form, $(D^2 - 2D + 1)y = e^x \log x$

$\therefore$ C. S. = C. F. + P. I.

i) To find C. F.

Its A. E. is $D^2 - 2D + 1 = 0$; $(D-1)^2 = 0$; $D = 1, \ 1$

$\therefore$ C. F. $= (C_1 + C_2 x)e^x = C_1 e^x + C_2 x e^x$

ii) To find P. I.

P. I. $= y_1 U + y_2 V$

Where, $U = -\int \frac{y_2}{W} X \, dx$, $V = \int \frac{y_1}{W} X \, dx$, $W = \begin{vmatrix} y_1 & y_2 \\ y_1' & y_2' \end{vmatrix}$

Now,   Compare   C. F. $= C_1 y_1 + C_2 y_2$

Here, $y_1 = e^x$; $y_2 = xe^x$

$y_1' = e^x$; $y_2' = xe^x + e^x$

$W = \begin{vmatrix} y_1 & y_2 \\ y_1' & y_2' \end{vmatrix} = \begin{vmatrix} e^x & xe^x \\ e^x & xe^x + e^x \end{vmatrix} = xe^{2x} + e^{2x} - xe^{2x}$; $W = e^{2x}$

$\therefore$ P. I. $= -y_1 \int \frac{y_2}{W} X \, dx + y_2 \int \frac{y_1}{W} X \, dx$

$$= -e^x \int \frac{xe^x}{e^{2x}} \cdot e^x \log x \, dx + x e^x \int \frac{e^x}{e^{2x}} e^x \log x \, dx$$

$$= -e^x \int x \log x \, dx + xe^x \int \log x \, dx$$

$$= -e^x \left[\log x \int x \, dx - \int \left(\frac{d}{dx} \log x \int x \, dx \right) dx \right]$$

$$+ xe^x \left[\log x \int 1 dx - \int \left(\frac{d}{dx} \log x \int 1 dx \right) dx \right]$$

$$= -e^x \left[\log x \frac{x^2}{2} - \int \frac{1}{x} \frac{x^2}{2} dx \right] + xe^x \left(x \log x - \int \frac{1}{x} x dx \right)$$

$$= -e^x \left[\frac{x^2}{2} \log x - \frac{1}{2} \frac{x^2}{2} \right] + xe^x (x \log x - x)$$

$$= -e^x \frac{x^2}{2} \log x + \frac{x^2}{4} e^x + x^2 e^x \log x - x^2 e^x$$

$$= \frac{1}{4}[-2e^x x^2 \log x + x^2 e^x + 4x^2 e^x \log x - 4x^2 e^x]$$

$$= \frac{1}{4}[2e^x x^2 \log x - 3x^2 e^x] \quad = \frac{1}{4} x^2 e^x (2 \log x - 3)$$

iii) $\therefore$ **C. S. is** $\quad y = (C_1 + C_2 x)e^x + \dfrac{1}{4} x^2 e^x (2 \log x - 3)$

Example 90: Solve by VOP $\left(D^2 - 1\right) y = e^{-x} \sin(e^{-x}) + \cos(e^{-x})$

Solution: Given by VOP $(D^2 - 1) y = e^{-x} \sin(e^{-x}) + \cos(e^{-x})$

C. S. = C. F. + P. I.

i) To find C. F.

Its A. E. is $D^2 - 1 = 0$; $D^2 = 1$; $D = \pm 1 = -1, \ 1$

$\therefore$ C. F. $= C_1 e^{-x} + C_2 e^x$

ii) To find P. I.

P. I. $= y_1 U + y_2 V$

Where, $\quad U = -\displaystyle\int \frac{y_2}{W} X \, dx$, $\quad V = \displaystyle\int \frac{y_1}{W} X \, dx$, $\quad W = \begin{vmatrix} y_1 & y_2 \\ y_1' & y_2' \end{vmatrix}$

Now, Compare C. F. $= C_1 y_1 + C_2 y_2$

Here, $y_1 = e^{-x}$; $y_2 = e^x$

$y_1' = -e^{-x}$; $y_2' = e^x$

$W = \begin{vmatrix} y_1 & y_2 \\ y_1' & y_2' \end{vmatrix} = \begin{vmatrix} e^{-x} & e^x \\ -e^{-x} & e^x \end{vmatrix} = e^x e^{-x} + e^x e^{-x} = 1 + 1 = 2$; $W = 2$

$\therefore$ P. I. $= -y_1 \displaystyle\int \frac{y_2}{W} X \, dx + y_2 \int \frac{y_1}{W} X \, dx$

$= -e^{-x} \displaystyle\int \frac{e^x}{2} [e^{-x} \sin(e^{-x}) + \cos(e^{-x})] dx + e^x \int \frac{e^{-x}}{2} [e^{-x} \sin(e^x) + \cos(e^{-x})] \, dx$

$= \dfrac{-e^{-x}}{2} \left[\displaystyle\int \sin(e^{-x}) + e^x \cos(e^{-x}) \right] dx + \frac{e^x}{2} \int [e^{-2x} \sin(e^{-x}) + \cos(e^{-x})] \, dx$

$= -\dfrac{e^{-x}}{2} \left[\displaystyle\int (\sin(e^{-x}) dx + \int e^x \cos(e^{-x}) \, dx \right]$

$\qquad + \dfrac{e^x}{2} \left[\displaystyle\int e^{-2x} \sin(e^{-x}) dx + \int e^{-x} \cos(e^{-x}) dx \right]$

$= \dfrac{-e^{-x}}{2} \left[\displaystyle\int \sin(e^{-x}) \, dx + \cos(e^{-x}) e^x - \int (-\sin(e^{-x})(-e^{-x}) e^x dx \right]$

$\qquad + \dfrac{e^x}{2} \left[-\displaystyle\int t \sin t \, dt + (-1) \int \cos t \, dt \right]$

$$= \frac{-e^{-x}}{2}\left[\int \sin(e^{-x}) \, dx + e^x \cos(e^{-x}) - \int \sin(e^{-x})dx\right]$$

$$+ \frac{e^x}{2}\{-[t(-\cos t) - (\sin t)] - \sin t\}$$

$$= \frac{-e^{-x}}{2}[e^{x \cdot} \cos(e^{-x})] + \frac{e^x}{2}[e^{-x} \cos(e^{-x}) - 2\sin(e^{-x})]$$

$$= -\frac{1}{2}\cos(e^{-x}) + \frac{1}{2}\cos(e^{-x}) - e^x \sin(e^{-x})$$

$$= - e^x \sin(e^{-x})$$

iii) ∴ C. S. is $y = C_1 e^{-x} + C_2 e^x - e^x \sin(e^{-x})$

Example 91: **Apply the method of variation of parameter to solve**
$(D^2 + 4)y = 4\sec 2x$

Solution: Given D. E. $(D^2 + 4)y = 4\sec 2x$

C. S. = C. F. + P. I.

i) To find C. F.

Its A. E. is $D^2 + 4 = 0$; $D^2 = -4$; $D = \pm 2i$

∴ C. F. = $C_1 \cos 2x + C_2 \sin 2x$

ii) To find P. I.

P. I. = $y_1 U + y_2 V$

Where, $U = -\int \frac{y_2}{W} X \, dx$, $V = \int \frac{y_1}{W} X \, dx$, $W = \begin{vmatrix} y_1 & y_2 \\ y_1' & y_2' \end{vmatrix}$

Now,   Compare C. F. = $C_1 y_1 + C_2 y_2$

Here, $y_1 = \cos 2x$; $y_2 = \sin 2x$

$y_1' = -2\sin 2x$; $y_2' = 2\cos 2x$

$W = \begin{vmatrix} y_1 & y_2 \\ y_1' & y_2' \end{vmatrix} = \begin{vmatrix} \cos 2x & \sin 2x \\ -2\sin 2x & 2\cos 2x \end{vmatrix}$

$= 2\cos^2 2x + 2\sin^2 2x = 2(\cos^2 2x + \sin^2 2x)$; $W = 2$

∴ P. I. = $-y_1 \int \frac{y_2}{W} X \, dx + y_2 \int \frac{y_1}{W} X \, dx$

$= -\cos 2x \int \frac{\sin 2x}{2} 4\sec 2x \, dx + \sin 2x \int \frac{\cos 2x}{2} 4\sec 2x \, dx$

$= \frac{-\cos 2x}{2} 4 \int \tan 2x \, dx + \frac{4}{2}\sin 2x \int 1 \, dx$

$= -2\cos 2x \frac{\log|\sec 2x|}{2} + 2\sin 2x \cdot x$

$= -\cos 2x \log|\sec 2x| + 2x \sin 2x$

iii) ∴ C. S. is $y = C_1 \cos 2x + C_2 \sin 2x - \cos 2x \log|\sec 2x| + 2x \sin 2x$

Example 92: Solve $\left(D^2 + 3D + 2\right) y = e^{e^x}$ by using VOP

Solution: Given D. E. is $\left(D^2 + 3D + 2\right) y = e^{e^x}$

  C. S. = C. F. + P. I.

i) To find C. F.

Its A. E. is $D^2 + 3D + 2 = 0$; $D = -1, \; -2$

∴ C. F. $= C_1 e^{-x} + C_2 e^{-2x}$

ii) To find P. I.

  P. I. $= y_1 U + y_2 V$

Where, $U = -\displaystyle\int \frac{y_2}{W} X \, dx$, $V = \displaystyle\int \frac{y_1}{W} X \, dx$, $W = \begin{vmatrix} y_1 & y_2 \\ y_1' & y_2' \end{vmatrix}$

Now,   Compare C. F. $= C_1 y_1 + C_2 y_2$

Here, $y_1 = e^{-x}$; $y_2 = e^{-2x}$

$y_1' = e^{-x}$; $y_2' = -2e^{-2x}$

$W = \begin{vmatrix} y_1 & y_2 \\ y_1' & y_2' \end{vmatrix} = \begin{vmatrix} e^{-x} & e^{-2x} \\ -e^{-x} & -2e^{-2x} \end{vmatrix} = -2e^{-3x} + e^{-3x}$; $W = -e^{-3x}$

∴   P. I. $= -y_1 \displaystyle\int \frac{y_2}{W} X \, dx + y_2 \displaystyle\int \frac{y_1}{W} X \, dx$

$= -e^{-x} \displaystyle\int \frac{e^{-2x}}{-e^{-3x}} e^{e^x} dx + e^{-2x} \displaystyle\int \frac{e^{-x}}{-e^{-3x}} e^{e^x} dx$

$= e^{-x} \displaystyle\int e^x . e^{e^x} dx - e^{-2x} \displaystyle\int e^{2x} e^{e^x} dx$

$= e^{-x} \displaystyle\int e^t dt - e^{-2x} \displaystyle\int t \, e^t \, dt$ $\{ \because \text{ put } e^x = t ; \; e^x dx = dt$

$= e^{-x} e^{e^x} - e^{-2x} (t \, e^t - e^t)$

$= e^{-x} e^{e^x} - e^{-2x} \left(e^x e^{e^x} - e^{e^x} \right) = e^{-x} e^{e^x} - e^{-x} e^{e^x} + e^{-2x} e^{e^x}$

$= e^{-2x} e^{e^x}$

iii) ∴ C. S. is $y = C_1 e^{-x} + C_2 e^{-2x} + e^{-2x} e^{e^x}$

Example 93: Solve $\left(D^2 + a^2\right) y = \sec ax$ by the method of VOP.

Solution: Given D. E. is $\left(D^2 + a^2\right) y = \sec ax$

∴ C. S. = C. F. + P. I.

i) To find C. F.

Its A. E. is $D^2 + a^2 = 0$; $D^2 = -a^2$; $D = \pm \, ia$

∴ C. F. $= C_1 \cos ax + C_2 \sin ax$

ii) To find P. I.

$\text{P.I.} = y_1 U + y_2 V$

Where, $\quad U = -\displaystyle\int \frac{y_2}{W} X\, dx,\quad V = \displaystyle\int \frac{y_1}{W} X\, dx,\quad W = \begin{vmatrix} y_1 & y_2 \\ y_1' & y_2' \end{vmatrix}$

Now, $\quad$ Compare $\text{C.F.} = C_1 y_1 + C_2 y_2$

Here, $\quad y_1 = \cos ax \quad ; \quad y_2 = \sin ax$

$y_1' = -a\sin ax \; ; \; y_2' = a\cos ax$

$W = \begin{vmatrix} y_1 & y_2 \\ y_1' & y_2' \end{vmatrix} = \begin{vmatrix} \cos ax & \sin ax \\ -a\sin ax & a\cos ax \end{vmatrix} = a\cos^2 ax + a\sin^2 ax \; ; \; W = a$

$\therefore \quad \text{P.I.} = -y_1 \displaystyle\int \frac{y_2}{W} X\, dx + y_2 \displaystyle\int \frac{y_1}{W} X\, dx$

$= -\cos ax \displaystyle\int \frac{\sin ax}{a}\sec ax\, dx + \sin ax \displaystyle\int \frac{\cos ax}{a}\sec ax\, dx$

$= \dfrac{-\cos ax}{a}\displaystyle\int \tan ax\, dx + \dfrac{\sin ax}{a}\displaystyle\int 1\, dx$

$= \dfrac{-\cos ax}{a}\dfrac{\log|\sec ax|}{a} + \dfrac{\sin ax}{a} x$

$= -\dfrac{1}{a^2}\cos ax \log(\sec ax) + \dfrac{x}{a}\sin ax$

$= \dfrac{1}{a^2}\cos ax \log|\cos ax| + \dfrac{x}{a}\sin ax$

iii) $\therefore$ **C. S. is** $\quad y = C_1 \cos ax + 2\, C_2 \sin ax + \cos ax \dfrac{\log|\cos ax|}{a^2} + \dfrac{x}{a}\sin ax$

Example 94: Solve by using VOP $(D^3 + D)y = \operatorname{cosec} ax$

Solution: Given D. E. $(D^3 + D)y = \operatorname{cosec} ax$

$\therefore$ C. S. = C.F. + P.I.

Let $Dy = p$

$\therefore D^3 y + Dy = \operatorname{cosec} x$

i. e. $D^2 . Dy + Dy = \operatorname{cosec} ax$

$D^2 p + p = \operatorname{cosec} x$

$(D^2 + 1)p = \operatorname{cosec} x$

i) **To find C. F.**

It's A. E. is $D^2 + 1 = 0; \quad D^2 = -1; \quad D^2 = \pm i$

$\quad$ C. F. = $C_1 \cos x + C_2 \sin x$

i) **To find P. I.**

$\text{P.I.} = y_1 U + y_2 V$

Where, $\quad U = -\displaystyle\int \frac{y_2}{W} X\, dx,\quad V = \displaystyle\int \frac{y_1}{W} X\, dx,\quad W = \begin{vmatrix} y_1 & y_2 \\ y_1' & y_2' \end{vmatrix}$

Now,   Compare C. F. $= C_1 y_1 + C_2 y_2$

Here, $y_1 = \cos x$; $y_2 = \sin x$

$y_1' = -\sin x$; $y_2' = \cos x$

$W = \begin{vmatrix} y_1 & y_2 \\ y_1' & y_2' \end{vmatrix} = \begin{vmatrix} \cos x & \sin x \\ -\sin x & \cos x \end{vmatrix} = \cos^2 x + \sin^2 x$; $W = 1$

$\therefore$ P. I. $= -\cos x \int \dfrac{\sin x}{1} \cdot \text{cosec} \, x \; dx + \sin x \int \dfrac{\cos x}{1} \text{cosec} \, x \; dx$

$= -\cos x \int 1 \, dx \; + \sin x \int \cot x \; dx$

$= -\cos x \, (x) + \sin x \log|\sin x|$

iii) $\therefore$ C. S. is

$Dy = p = C_1 \cos x + C_2 \sin x - x \cos x + \sin x \log|\sin x|$

Integreting bothsides, we get

$y = C_1 \sin x - C_2 \cos x + C_3 - [x \sin x - (-\cos x)] + \log|\sin x| \, (-\cos x)$

$\quad - \int \dfrac{1}{\sin x} \cos x \, (-\cos x) \, dx$

$= C_1 \sin x - C_2 \cos x + C_3 - x \sin x - \cos x - \cos x \log|\sin x| + \int \dfrac{1 - \sin^2 x}{\sin x} \, dx$

$= C_1 \sin x - C_2 \cos x + C_3 - x \sin x - \cos x - \cos x \log|\sin x|$
$\quad + \log|\text{cosec} \, x - \cot x| + \cos x$

$y = C_1 \sin x - C_2 \cos x + C_3 - x \sin x - \cos x \log|\sin x| + \log|\text{cosec} \, x - \cot x|$

[Note: This problem solve by Alternate method see problem number (96)]

Example 95: Solve by VOP $x^2 \dfrac{d^2 y}{dx^2} + x \dfrac{dy}{dx} - y = \dfrac{x^3}{1 + x^2}$

Solution: Given D. E. $x^2 \dfrac{d^2 y}{dx^2} + x \dfrac{dy}{dx} - y = \dfrac{x^3}{1 + x^2}$ ...... (1)

This is cauchy's D. E.

$\therefore$ Put $x = e^z$, $z = \log x$, $\dfrac{d}{dz} = D$

$\therefore x^2 \dfrac{d^2 y}{dx^2} = D(D-1)y$; $x \dfrac{dy}{dx} = Dy$

$\therefore$ Equation (1)$\rightarrow [D(D-1) + D - 1]y = \dfrac{e^{3z}}{1 + e^{2z}}$

$(D^2 - 1)y = \dfrac{e^{3z}}{1 + e^{2z}}$ ... Which is L. D. E. with constant coefficient

$\therefore$ C. S. $=$ C. F. + P. I.

i) **To find C. F.**

Its A. E. is $D^2 - 1 = 0$; $D^2 = 1$; $D = \pm 1 = -1, 1$

$\therefore$ C. F. $= C_1 e^{-z} + C_2 e^z = C_1 x^{-1} + C_2 x$

ii) To find P. I.

P. I. $= y_1 U + y_2 V$

Where, $U = -\displaystyle\int \frac{y_2}{W} X \, dx$, $V = \displaystyle\int \frac{y_1}{W} X \, dx$, $W = \begin{vmatrix} y_1 & y_2 \\ y_1' & y_2' \end{vmatrix}$

Now,   Compare   C. F. $= C_1 y_1 + C_2 y_2$

Here, $y_1 = e^{-z}$; $y_2 = e^z$

$y_1' = -e^{-z}$; $y_2' = e^z$

$W = \begin{vmatrix} y_1 & y_2 \\ y_1' & y_2' \end{vmatrix} = \begin{vmatrix} e^{-z} & e^z \\ -e^{-z} & e^z \end{vmatrix} = e^z e^{-z} + e^z e^{-z} = 1 + 1$; $W = 2$

$\therefore$ P. I. $= -y_1 \displaystyle\int \frac{y_2}{W} X \, dx + y_2 \int \frac{y_1}{W} X \, dx$

$= -e^{-z} \displaystyle\int \frac{e^z}{2} \frac{e^{3z}}{1 + e^{2z}} \, dz + e^z \int \frac{e^{-z}}{2} \frac{e^{3z}}{1 + e^{2z}} \, dz$

$= \dfrac{-e^{-z}}{2} \displaystyle\int \frac{e^{4z}}{1 + e^{2z}} \, dz + \frac{e^z}{2} \int \frac{e^{2z}}{1 + e^{2z}} \, dz$

$= \dfrac{-e^{-z}}{2} \displaystyle\int \frac{(t-1)\frac{dt}{2}}{t} + \frac{e^z}{2} \int \frac{\frac{dt}{2}}{t} \begin{cases} \because \text{Put } 1 + e^{2z} = t; \ e^{2z} = t - 1 \\ 2e^{2z} \, dz = dt; \ e^{2z} \, dz = 1/2 \, dt \end{cases}$

$= \dfrac{-e^{-z}}{2.2} \displaystyle\int \frac{(t-1)}{t} \, dt + \frac{e^z}{2.2} \int \frac{1}{t} \, dt$

$= \dfrac{-e^{-z}}{4} \displaystyle\int \left(1 - \frac{1}{t}\right) dt + \frac{e^z}{4} \log t$

$= \dfrac{-e^z}{4}(t - \log t) + \frac{e^z}{4} \log t$

$= -\dfrac{e^{-z}}{4}[(1 + e^{2z}) - \log(1 + e^{2z})] + \frac{e^z}{4} \log(1 + e^{2z})$

$= -\dfrac{e^{-z}}{4}(1 + e^{2z}) + \frac{e^{-z}}{4} \log(1 + e^{2z}) + \frac{e^z}{4} \log(1 + e^{2z})$

$= -\dfrac{1}{4}(e^{-z} + e^z) + \frac{1}{4} \log(1 + e^{2z})[e^z + e^{-z}]$

$= -\dfrac{1}{2} \cosh z + \frac{1}{4} \log(1 + e^{2z}) \, 2\sinh z$

$= -\dfrac{1}{2} \cosh(\log x) + \frac{1}{2} \log(1 + x^2) \sinh(\log x)$

iii) $\therefore$ **C. S. is** $y = C_1 x^{-1} + C_2 x - \dfrac{1}{2}[\cosh(\log x) + \log(1 + x^2) \sinh(\log x)]$

Example 96: Solve $(D^3 + D)y = \csc x$ using method of VOP

Solution: Given D. E. $(D^3 + D) = \csc x$

$\therefore$ C. S. $=$ C. F. $+$ P. I.

i) To find C. F.

Its A. E. is $D^3 + D = 0$; $D(D^2 + 1) = 0$; $D = 0$, $D^2 = -1$

$\therefore D = 0, \pm i$

$\therefore$ C. F. $= C_1 + C_2 \cos x + C_3 \sin x$

ii) To find P. I.

$\therefore$ P. I. $= y_1 U + y_2 V + y_3 M$

Where, $U = \int \dfrac{\begin{vmatrix} y_2 & y_3 \\ y_2' & y_3' \end{vmatrix}}{W} X \, dx$, $V = -\int \dfrac{\begin{vmatrix} y_1 & y_3 \\ y_1' & y_3' \end{vmatrix}}{W} X \, dx$, $M = \int \dfrac{\begin{vmatrix} y_1 & y_2 \\ y_1' & y_2' \end{vmatrix}}{W} X \, dx$

Now,   Compare C. F. $= C_1 y_1 + C_2 y_2 + C_3 y_3$

Here, $y_1 = 1$; $y_2 = \cos x$; $y_3 = \sin x$

$y_1' = 0$; $y_2' = -\sin x$; $y_3' = \cos x$

$y_1'' = 0$; $y_2'' = -\cos x$; $y_3'' = -\sin x$

$W = \begin{vmatrix} y_1 & y_2 & y_3 \\ y_1' & y_2' & y_3' \\ y_1'' & y_2'' & y_3'' \end{vmatrix} = \begin{vmatrix} 1 & \cos x & \sin x \\ 0 & -\sin x & \cos x \\ 0 & -\cos x & -\sin x \end{vmatrix} = \cos^2 x + \sin^2 x = 1$

$\therefore W = 1$

$\therefore$ P. I. $= y_1 \int \dfrac{\begin{vmatrix} y_2 & y_3 \\ y_2' & y_3' \end{vmatrix}}{W} X \, dx - y_2 \int \dfrac{\begin{vmatrix} y_1 & y_3 \\ y_1' & y_3' \end{vmatrix}}{W} X \, dx + y_3 \int \dfrac{\begin{vmatrix} y_1 & y_2 \\ y_1' & y_2' \end{vmatrix}}{W} X \, dx$

$= 1 \int \dfrac{\begin{vmatrix} \cos x & \sin x \\ -\sin x & \cos x \end{vmatrix}}{1} \csc x \, dx - \cos x \int \dfrac{\begin{vmatrix} 1 & \sin x \\ 0 & \cos x \end{vmatrix}}{1} \csc dx + \sin x \int \dfrac{\begin{vmatrix} 1 & \cos x \\ 0 & -\sin x \end{vmatrix}}{1} \csc dx$

$= \int (\cos^2 x + \sin^2 x) \csc x \, dx - \cos x \int (\cos x) \csc x \, dx + \sin x \int (-\sin x) \csc x \, dx$

$= \int \csc x \, dx - \cos x \int \cot x \, dx - \sin x \int 1 dx$

$= \log|\csc x - \cot x| - \cos x \log|\sin x| - \sin x \, (x)$

iii) $\therefore$ **C. S. is**

$y = C_1 + C_2 \cos x + C_3 \sin x + \log|\csc x - \cot x| - \cos x \log|\sin x| - \sin x \, (x)$

8. ii   Method of undermined coefficients by VOP

A] Steps to solve for 2nd order LDE	B] Steps to solve for 3rd order LDE
i) Let the given D. E.	i) Let the given D. E.
$$\frac{d^2y}{dx^2} + k_1 \frac{dy}{dx} + k_2 y = X$$	$$\frac{d^3y}{dx^3} + k_1 \frac{d^2y}{dx^2} + k_2 \frac{dy}{dx} + k_3 y = X$$
Where k's are constant and X is f(x)	Where k's are constant and X is f(x)
ii) Let C. F. $= C_1 y_1 + C_2 y_2$	ii) Let C. F. $= C_1 y_1 + C_2 y_2 + C_3 y_3$
iii) P. I. $= U\, y_1 + V\, y_2$	iii) P. I. $= U\, y_1 + V\, y_2 + W\, y_3$
Where U &V one unknown functions of x satisfying	WhereU, V, &W are unknown functions of x satisfying
$U' y_1 + V' y_2 = 0$	$U' y_1 + V' y_2 + W' y_3 = 0$
$U' y_1' + V' y_2' = X$	$U' y_1' + V' y_2' + W' y_3' = 0$
Solve above equation for U', V' and then	$U' y_1'' + V' y_2'' + W' y_3'' = X$
integration wrt. x, we get U &V so that	iv) Solve above equation for $U', V' \& W'$
iv) P. I. $= U y_1 + V y_2$	then integrating w. r. t. x
$\therefore$ C. S. $=$ C. F. $+$ P. I.	we get U, V & W so that
	P. I. $= U y_1 + V y_2 + W y_3$
	$\therefore$ C. S. $=$ C. F. $+$ P. I.

8. ii. a   Examples on Method of undermined coefficients by VOP

Example 97: Using VOP solve $\dfrac{d^2y}{dx^2} + a^2 y = \sec ax$

Solution: Given D. E. in symbolic form $(D^2 + a^2)y = \sec ax$

$\therefore$ C. F. $=$ C. F. $+$ P. I.

i) To find C. F.

Its A. E. is $D^2 + a^2 = 0$; $D^2 = -a^2$; $D = \pm\, ia$

$\therefore$ C. F. $= C_1 \cos ax + C_2 \sin ax$

ii) To find P. I.

   P. I. $= y_1 U + y_2 V$

Now,   Compare C. F. $= C_1 y_1 + C_2 y_2$

Here, $y_1 = \cos ax$; $y_2 = \sin ax$

$y_1' = -a \sin ax$; $y_2' = a \cos ax$

Where,   U and V are unknown function of x satisfying

$U' y_1 + V' y_2 = 0$; $U' \cos ax + V' \sin ax = 0$ ... ... (i)

$U' y_1' + V' y_2' = X$; $U' (-a \sin ax) + V'a \cos ax = \sec ax$ ... ... (ii)

Now,

Equn(i) $\times$ a cos ax U' a $\cos^2$ ax $+$ V' a cos ax . sin ax $= 0$

Equn(ii) $\times$ sin ax $- U'$ a $\sin^2$ ax $+ V'$ a cos ax . sin ax $=$ tan ax

Subtracting $+$ $-$ $-$

$$U' a = - \tan ax$$

$$\therefore \quad U' = \frac{-\tan ax}{a}$$

Put in equn(i), $\dfrac{-\tan ax}{a} \cos ax + V'\sin ax = 0$

$\dfrac{-\sin ax}{a} + V'\sin ax = 0$

$V'\sin ax = \dfrac{\sin ax}{a}$

$$V' = \frac{1}{a}$$

Now,   Integrating U' & V'

$\therefore U = \displaystyle\int U'dx = \dfrac{-1}{a}\int \tan ax\, dx = \dfrac{-1}{a^2}\log|\sec ax| = \dfrac{1}{a^2}\log|\cos ax|$

$V = \displaystyle\int V'dx = \dfrac{1}{a}\int 1dx = \dfrac{x}{a}$

$\therefore$ P. I. $= \dfrac{1}{a^2}\log|\cos ax| . \cos ax + \dfrac{x}{a}\sin ax$

iii) $\therefore$ **C. S. is** $y = C_1 \cos ax + C_2 \sin ax + \dfrac{1}{a^2}\log|\cos ax| . \cos ax + \dfrac{1}{a} x \sin ax$

Example 98: Using VOP solve $(D^2 + 4)y = 4\sec^2 2x$

Solution: Given D. E. is $(D^2 + 4)y = 4\sec^2 2x$

  C. S. $=$ C. F. $+$ P. I.

i) To find C. F.

Its A. E. is $D^2 + 4 = 0$: $D^2 = -4$; $D = \pm 2i$

$\therefore$ C. F. $= C_1 \cos 2x + C_2 \sin 2x$

ii) To find P. I.

  P. I. $= y_1 U + y_2 V$

Now,   Compare   C. F. $= C_1 y_1 + C_2 y_2$

Here, $y_1 = \cos 2x$; $y_2 = \sin 2x$

$y_1' = -2 \sin 2x$; $y_2' = 2 \cos 2x$

Where,   U and V are unknown function of x satisfying

$U'y_1 + V'y_2 = 0$; $U'\cos 2x + V'\sin 2x = 0$ ... ... (i)

$U'y_1' + V'y_2' = X$; $U'(-2\sin 2x) + V'(2\cos 2x) = 4\sec^2 2x$ ... ... (ii)

Equn(i) $\times$ 2 cos 2x 2U$'$ cos^2 2x + V$'$ 2 cos 2x . sin 2x = 0

Equn(ii) $\times$ sin 2x $-$ 2U$'$ sin^2 2x + V$'$ 2 cos 2x . sin 2x = 4 sin 2x . sec^2 2x

Subtracting   + $-$ $-$

$$\overline{}$$

$$2U' = -4 \sin 2x . \sec^2 2x$$

$$U' = \frac{-4 \sin 2x . \sec^2 2x}{2}$$

$\therefore$ U$'$ = -2 sec 2x . tan 2x

Put in equn (i) $-$ 2 sec 2x . tan 2x . cos 2x + V$'$ sin 2x = 0

$$-2 \tan 2x + V' \sin 2x = 0$$

$$V' = \frac{2 \tan 2x}{\sin 2x} = 2 \sec 2x$$

$\therefore$ V$'$ = 2 sec 2x

Now,   Integrating U$'$ & V$'$

$$\therefore U = \int U' dx = \int -2 \sec 2x . \tan 2x \ dx = -2 \frac{\sec 2x}{2} = -\sec 2x$$

$$V = \int V' dx = \int 2 \sec 2x \ dx = 2 \frac{\log|\sec 2x + \tan 2x|}{2} = \log|\sec 2x + \tan 2x|$$

$\therefore$ P. I. = $-$ cos 2x sec 2x + sin 2x log|sec 2x + tan 2x|

iii) $\therefore$ **C. S. is**   **y = C$_1$ cos 2x + C$_2$ sin 2x $-$ 1 + sin 2x log|sec 2x + tan 2x|**

Example 99: Evaluate $\dfrac{d^2y}{dx^2} + y = \text{cosec } x,$ **using the method of VOP**

Solution: Given D. E. in symbolic form $(D^2 + 1)y = \text{cosec } x$

   C. S. = C. F. + P. I.

i) To find C. F.

Its A. E. is $D^2 + 1 = 0$; $D^2 = -1$;   D = $\pm$i

$\therefore$ C. F. = $C_1 \cos x + C_2 \sin x$

ii) To find P. I.

   P. I. = $y_1 U + y_2 V$

Now,   Compare   C. F. = $C_1 y_1 + C_2 y_2$

Here, $y_1 = \cos x$; $y_2 = \sin x$

$y_1' = -\sin x$; $y_2' = \cos x$

Where,   U and V are unknown function of x satisfying

U$'$y$_1$ + V$'$y$_2$ = 0 ;   U$'$ cos 2x + V$'$ sin 2x = 0   ... ... (i)

U$'$y$_1'$ + V$'$y$_2'$ = X ; $-$U$'$ sin x + V$'$ cos x = cosec x   ... ... (ii)

Equn(i) $\times$ cos x U$'$ cos^2 x + V$'$ sin x . cos x = 0

Equn(ii) $\times \sin x$ $-U' \sin^2 x + V' \sin x. \cos x = \sin x. \operatorname{cosec} x$

Subtracting $+$ $-$ $-$

$U' = -1$

Put in equn equn(i), $-\cos x + V' \sin x = 0$

$V' \sin x = \cos x = \dfrac{\cos x}{\sin x}$

$V' = \cot x$

Now, Integrating $U' \& V'$

$U = \displaystyle\int U' \, dx = \int -1 \, dx = -x$

$V = \displaystyle\int V' \, dx = \int \cot x \, dx = \log|\sin x|$

$\therefore$ P.I. $= -x \cos x + \sin x \log|\sin x|$

iii) $\therefore$ **C.S. is** $y = C_1 \cos x + C_2 \sin x - x \cos x + \sin x \log|\sin x|$

Eample 100: Solve $\dfrac{dy}{dx} - y = 3e^{-x}$ by VOP

Solution: Given D.E. in symbolic form $(D-1)y = 3e^{-x}$

   $\therefore$ C.S. $=$ C.F. $+$ P.I.

i) To find C.F.

Its A.E. is $D - 1 = 0$; $D = 1$.

$\therefore$ C.F. $= C_1 e^x$

ii) To find P.I.

   P.I. $= y_1 U$

Now,   Compare C.F. $= C_1 y_1$

Here, $y_1 = e^x$

Where,   U is unknown function of x satisfying

$U' y_1 = X$; $U' e^x = 3e^{-x}$

$U' = 3e^{-2x}$

$\therefore$ Integrating, $U = \displaystyle\int U' dx = 3 \int e^{-2x} dx = 3\dfrac{e^{-2x}}{-2} = \dfrac{-3e^{-2x}}{2}$

$\therefore$ P.I. $= y_1 U = e^x \left(\dfrac{-3e^{-2x}}{2}\right) = \dfrac{-3}{2} e^{-x}$

iii) $\therefore$ **C.S. is** $y = C_1 e^x - \dfrac{3}{2} e^{-x}$

Example 101: Solve $(D^3 + D)y = \operatorname{cosec} x$ using method of VOP

Solution: Given D.E. $(D^3 + D)y = \operatorname{cosec} x$

C. S. = C. F. +P. I.

i) To find C. F.

Its A. E. is $D^3 + D = 0$; $D(D^2 + 1) = 0$; $D = 0$, $D^2 = -1$; $\therefore D = 0$, $\pm i$

$\therefore$ C. F. $= C_1 + C_2 \cos x + C_3 \sin x$

ii) To find P. I.

P. I. $=$ U y_1 + V y_2 + W y_3

Now,   Compare   C. F. $= C_1 y_1 + C_2 y_2 + C_3 y_3$

Here, $y_1 = 1$; $y_2 = \cos x$; $y_3 = \sin x$

$y_1' = 0$; $y_2' = -\sin x$; $y_3' = \cos x$

$y_1'' = 0$; $y_2'' = -\cos x$; $y_3'' = -\sin x$

Where U, V and W are unknown functions of x such that

$U' y_1 + V' y_2 + W' y_3 = 0$; $U' + V' \cos x + W' \sin x = 0$ ... ... (i)

$U' y_1' + V' y_2' + W' y_3' = 0$; $-V' \sin x + W' \cos x = 0$ ... ... (ii)

$U' y_1'' + V' y_2'' + W' y_3'' = X$; $-V' \cos x - W' \sin x = \csc x$ ... ... (iii)

Equn(ii) $\times \sin x$ $-V' \sin^2 x + W' \sin x . \cos x = 0$

Equn(iii) $\times \cos x$ $-V' \cos^2 x - W' \sin x . \cos x = \cos x . \csc x$

Adding $\overline{\qquad\qquad\qquad -V' \quad = \quad \cot x \qquad}$

$$\therefore V' = - \cot x$$

Put in equn (ii) $\rightarrow$ $-(-\cot x) \sin x + W' \cos x = 0$

$W' = -1$

Put in equn(i) $\rightarrow$ $U' + (-\cot x) . \cos x + (-1) \sin x = 0$

$U' = \dfrac{\cos x}{\sin x} \cos x + \sin x = \dfrac{\cos^2 x + \sin^2 x}{\sin x}$

$U' = \csc x$

Now,   Integrating U', V' & W'

$$U = \int U' dx = \int \csc x \, dx = \log(\csc x - \cot x)$$

$$V = \int V' dx = -\int \cot x \, dx = - \log(\sin x)$$

$$W = \int W' dx = -\int 1 dx = -x$$

$\therefore$ P. I. $= \log(\csc x - \cot x) - \cos x \log(\sin x) - x \sin x$

iii) $\therefore$ C. S. is

$y = C_1 + C_2 \cos x + C_3 \sin x + \log(\csc x - \cot x) - \cos x . \log(\sin x) - x \sin x$

9   Simultaneous Differential Equations

Quite after we come across linear differential equations in which there are two or more depended variables and a single independent variable. Such equations are known as simultaneous linear equations.

Here we shall deal with systems of linear equations with constant coefficient only such a system of equations is solved by eliminating all but one of the dependent variables and then solving the resulting equations as before. Each of the dependent variables is obtained in a similar manner.

9.i   Examples on Simultaneous Differential Equations

Example 102: Solve the simultaneos equations:

$$\frac{dx}{dt} - wy = a\cos pt \quad ; \quad \frac{dy}{dt} + wx = a\sin pt$$

Solution: Given equation in symbolic form

$$Dx - wy = a\cos pt \qquad \qquad \text{...... (i)}$$
$$wx + Dy = a\sin pt \qquad \qquad \text{...... (ii)}$$

Equn(i) × w $wDx - w^2y = aw\cos pt$

Equn(ii) × D $Dwx + D^2y = aD\sin pt$

Subtracting $\underline{\quad - \qquad - \qquad - \qquad\qquad\qquad}$

$$-(D^2 + w^2)y = aw\cos pt - ap\cos pt$$

$$\left(D^2 + w^2\right)y = a(p - w)\cos pt \qquad \text{... Which is LDE with constant coefficient}$$

∴ C.S. = C.F. + P.I.

i) To find C. F.

Its A. E. is $D^2 + w^2 = 0$; $D^2 = -w^2$; $D = \pm iw$

∴ C. F. = $C_1\cos wt + C_2\sin wt$

ii) To find P.I.

$$\therefore \text{P. I.} = \frac{1}{D^2 + w^2}\, a\,(p - w)\cos pt$$

$$= a\,(p - w)\frac{\cos pt}{(-1)p^2 + w^2} \quad = \frac{a(p - w)\cos pt}{-(p^2 - w^2)} \quad = \frac{a(p - w)\cos pt}{-(p - w)(p + w)}$$

$$\text{P. I.} = \frac{-a\cos pt}{p + w}$$

iii) ∴ **C.S. is** $y = C_1\cos wt + C_2\sin wt - \dfrac{a\cos pt}{p + w}$ ...... (iii)

From equn(ii) becomes, $wx = a\sin pt - Dy$

$$= a\sin pt - \frac{d}{dt}\left[C_1\cos wt + C_2\sin wt - \frac{a\cos pt}{p + w}\right]$$

$$= a \sin pt + C_1 w \sin wt - C_2 w \cos wt - \frac{a}{p+w} p \sin pt$$

$$= w C_1 \sin wt - w C_2 \cos wt + a \left(1 - \frac{p}{p+w}\right) \sin pt$$

$$\therefore \quad x = C_1 \sin wt - C_2 \cos wt + \frac{a}{p+w} \sin pt \qquad \text{...... (iv)}$$

Thus the equn (iii) and (iv) consititute the solution of equations (i) & (ii)

Example 103: Solve the simultaneous equation

$$\frac{dx}{dt} + 5x - 2y = t \; ; \; \frac{dy}{dt} + 2x + y = 0. \text{ Being given } x = y = 0 \; ; \text{ when } t = 0.$$

Solution: Given D. E. in symbolic form

$$Dx + 5x - 2y = t \; ; \quad (D+5)x - 2y = t \qquad \text{...... (i)}$$

$$Dy + 2x + y = 0 \; ; \quad 2x + (D+1)y = 0 \qquad \text{...... (ii)}$$

Now,

Equn: (i) × 2	$2(D+5)x - 4y$	$= 2t$
Equn: (ii) × (D + 5)	$2(D+5)x + (D+5)(D+1)y$	$= 0$
Subtracting	$- \qquad\qquad -$	$-$

$$-4y - (D^2 + 6D + 5)y = 2t$$

$$-(D^2 + 6D + 9)y = 2t$$

$$\left(D^2 + 6D + 9\right)y = -2t \qquad \text{... Which is L. D. E. with constant coefficient}$$

$$C. S. = C. F. + P. I.$$

i) To find C. F.

Its A. E. is $(D^2 + 6D + 9)y = 0; \quad (D+3)^2 = 0 \; ; \; D = -3, -3.$

$$\therefore C. F. = (C_1 + C_2 t)e^{-3t}$$

ii) To find P. I.

$$\therefore \quad P. I. = \frac{1}{(D+3)^2}(-2t)$$

$$= -2 \frac{1}{3^2 \left(1 + \frac{D}{3}\right)^2} t \qquad = \frac{-2}{9} \left(1 + \frac{D}{3}\right)^{-2} t$$

$$= \frac{-2}{9} \left(1 - \frac{2D}{3} + \cdots\right) t \qquad = \frac{-2}{9} \left(t - \frac{2}{3}\right)$$

iii) $\therefore$ **C. S. is** $\quad y = (C_1 + C_2 t)e^{-3t} - \frac{2}{9}\left(t - \frac{2}{3}\right) \text{...... (iii)}$

Now, Equn(ii) becomes,

$$2x = -(D+1)\left[(C_1 + C_2 t)e^{-3t} - \frac{2}{9}\left(t - \frac{2}{3}\right)\right]$$

$$= -(C_1 + C_2 t)(-3)e^{-3t} - e^{-3t}(C_2) + \frac{2}{9} - (C_1 + C_2 t)e^{-3t} + \frac{2}{9}\left(t - \frac{2}{3}\right)$$

$$= 3C_1 e^{-3t} + 3C_2 t\, e^{-3t} - C_2 e^{-3t} + \frac{2}{9} - C_1 e^{-3t} - C_2 t\, e^{-3t} + \frac{2t}{9} - \frac{4}{27}$$

$$= (3C_1 + 3C_2 - C_2 - C_1 - C_2 t)e^{-3t} + \frac{2t}{9} + \frac{2}{9} - \frac{4}{27}$$

$$2x = (2C_1 + 2C_2 t - C_2)e^{-3t} + \frac{2t}{9} + \frac{2}{27}$$

$$\therefore x = \left(C_1 + C_2 t - \frac{C_2}{2}\right)e^{-3t} + \frac{t}{9} + \frac{1}{27} \quad \text{...... (iv)}$$

$\therefore$ Equn (iii) and equn(iv) are the solution of equn (i) & (ii)

Now,   Given x = 0, y = 0 when t = 0

$\therefore$ Equn (iii) $0 = C_1 + \dfrac{4}{27}$; $\quad C_1 = \dfrac{-4}{27}$

Equn (iv) $0 = C_1 - \dfrac{C_2}{2} + \dfrac{1}{27}$; $\dfrac{C_2}{2} = \dfrac{-4}{27} + \dfrac{1}{27}$; $\dfrac{C_2}{2} = \dfrac{-3}{27}$; $C_2 = \dfrac{-6}{27}$

$\therefore$ Equn (iii) becomes, $y = \left(\dfrac{-4}{27} - \dfrac{6}{27} t\right)e^{-3t} - \dfrac{2}{9}\left(t - \dfrac{2}{3}\right)$

$\therefore \quad y = \dfrac{-2}{27}(2 + 3t)e^{-3t} - \dfrac{2t}{9} + \dfrac{4}{27}$

Equn(iv) becomes, $x = \left[\dfrac{-4}{27} + \dfrac{(-6)}{27} t - \left(\dfrac{-6}{27}\right)\dfrac{1}{2}\right]e^{-3t} + \dfrac{t}{9} + \dfrac{1}{27}$

$$\therefore x = \dfrac{-1}{27}(1 + 6t)e^{-3t} + \dfrac{t}{9} + \dfrac{1}{27}$$

Example 104: Solve the simultaneous equations $\dfrac{dx}{dt} + 2y + \sin t = 0$;

$\dfrac{dy}{dt} - 2x - \cos t = 0$ given that x = 0 and y = 1 when t = 0

Solution: Given D. E. in symbolic form

$\qquad Dx + 2y + \sin t = 0$; $Dx + 2y = -\sin t$ ...... (i)

$\qquad Dy - 2x - \cos t = 0$; $-2x + Dy = \cos t$ ...... (ii)

Equn(i) $\times -2$ $-2Dx - 4y = 2\sin t$

Equn(ii) $\times$ D $-2Dx + D^2 y = D\cos t$

Subtracting $\overline{\quad - \qquad - \qquad - \qquad\quad}$

$\qquad\qquad\qquad -4y - D^2 y = 2\sin t - D\cos t$

$\qquad\qquad\qquad -(D^2 + 4)y = 2\sin t + \sin t$

$(D^2 + 4)y = -3\sin t$ ... Which is L. D. E. with constant coefficient

C.S. = C.F. + P.I.

i) **To find C. F.**

Its A. E. is $D^2 + 4 = 0$; $D^2 = -4$; $D = \pm 2i$

$\therefore$ C. F. $= C_1 \cos 2t + C_2 \sin 2t$

ii) **To find P. I.**

$$\text{P. I.} = \frac{1}{D^2 + 4}(-3 \sin t)$$

$$= -3\frac{1}{D^2 + 4}\sin t \quad = \frac{-3 \sin t}{(-1)(1)^2 + 4} \quad = \frac{-3 \sin t}{3} \quad = -\sin t$$

iii) $\therefore$ **C. S. is** $y = C_1 \cos 2t + C_2 \sin 2t - \sin t$ ... ... (iii)

Put in equn(ii), $2x = Dy - \cos t$

$2x = D(C_1 \cos 2t + C_2 \sin 2t - \sin t) - \cos t$

$2x = -2C_1 \sin 2t + 2C_2 \cos 2t - \cos t - \cos t$

$2x = -2[C_1 \sin 2t - C_2 \cos 2t + \cos t]$

$\therefore x = -(C_1 \sin 2t - C_2 \cos 2t + \cos t)$ ... ... (iv)

Now,   Given $x = 0$, $y = 1$ when $t = 0$

Equn (iii)$\rightarrow$ $1 = C_2$

Equn (iv) $\rightarrow$ $0 = -C_2 + 1$; $C_2 = 1$

$\therefore$ $y = \cos 2t + \sin 2t - \sin t$ } **Required solution.**
    $x = \cos 2t + \sin 2t - \cos t$

Example 105: Solve the simultaneous equations:

$\dfrac{dx}{dt} + \dfrac{dy}{dt} - 2y = 2\cos t - 7\sin t$, $\dfrac{dx}{dt} - \dfrac{dy}{dt} + 2x = 4\cos t - 3\sin t$

Solution: Given D. E. in symbolic form

$Dx + Dy - 2y = 2 \cos t - 7 \sin t$; $Dx - Dy + 2x = 4 \cos t - 3 \sin t$

i. e. $Dx + (D - 2)y = 2 \cos t - 7 \sin t$ ... ... (i)

$(D + 2)x - Dy = 4 \cos t - 3\sin t$ ... ... (ii)

Equn(i) $\times$ $(D + 2)$ $(D + 2)Dx + (D + 2)(D - 2)y = (D + 2)[2\cos t - 7\sin t]$

Equn(ii) $\times$ D $(D + 2)Dx - D^2 y$ $= D[4 \cos t - 3 \sin t]$

Subtracting $-$ $+$ $-$

$(D^2 - 4 + D^2)y = -2\sin t - 7\cos t + 4\cos t - 14\sin t + 4 \sin t + 3\cos t$

i. e. $(2D^2 - 4)y = -12\sin t$

$(D^2 - 2)y = -6 \sin t$ ... Which is L. D. E. with constant coefficient

$\therefore$ C. S. $=$ C. F. + P. I.

i) **To find C. F.**

Its A. E. is $D^2 - 2 = 0$; $D^2 = 2$; $D = \pm\sqrt{2} = \sqrt{2}, -\sqrt{2}$

$\therefore$ C. F. $= C_1 e^{\sqrt{2}\,t} + C_2 e^{-\sqrt{2}\,t}$

ii) To find P. I.

$$P. I. = \frac{1}{D^2 - 2}(-6\sin t)$$

$$= -6\frac{1}{D^2 - 2}\sin t \qquad = -6\frac{1}{(-1)(1)^2 - 2}\sin t \qquad = \frac{-6\sin t}{-3}$$

P. I. $= 2\sin t$

iii) $\therefore$ **C. S. is** $y = C_1 e^{\sqrt{2}\,t} + C_2 e^{-\sqrt{2}\,t} + 2\sin t$

Now, equn(i)$\rightarrow$ $Dx = -(D - 2)y + 2\cos t - 7\sin t$

$Dx = -(D - 2)\left[C_1 e^{\sqrt{2}\,t} + C_2 e^{-\sqrt{2}\,t} + 2\sin t\right] + 2\cos t - 7\sin t$

$Dx = -DC_1 e^{\sqrt{2}\,t} - DC_2 e^{-\sqrt{2}\,t} - D2\sin t + 2C_1 e^{\sqrt{2}\,t} + 2C_2 e^{-\sqrt{2}\,t} + 4\sin t + 2\cos t - 7\sin t$

$Dx = -\sqrt{2}\,C_1 e^{\sqrt{2}\,t} + \sqrt{2}\,C_2 e^{-\sqrt{2}\,t} - 2\cos t + 2\sqrt{2}\,C_1 e^{\sqrt{2}\,t} - 2\sqrt{2}\,C_2 e^{-\sqrt{2}\,t} + 2\cos t$

     $- 3\sin t$

$= (-\sqrt{2} + 2\sqrt{2})C_1 e^{\sqrt{2}\,t} + (\sqrt{2} - 2\sqrt{2})C_2 e^{-\sqrt{2}\,t} + 2\cos t - 3\sin t$

$= (\sqrt{2})C_1 e^{\sqrt{2}\,t} + (-\sqrt{2})C_2 e^{-\sqrt{2}\,t} + 2\cos t - 3\sin t$

Integrating both sides

$\therefore$ $x = \int Dx\, dx = \sqrt{2}C_1 \int e^{\sqrt{2}\,t}\, dt - \sqrt{2}C_2 \int e^{-\sqrt{2}\,t}\, dt + 2\int \cos t\, dt - 3\int \sin t\, dt$

$$= \sqrt{2}C_1 \frac{e^{\sqrt{2}\,t}}{\sqrt{2}} - \sqrt{2}C_2 \frac{e^{-\sqrt{2}\,t}}{-\sqrt{2}} + 2\sin t + 3\cos t$$

$x = C_1 e^{\sqrt{2}\,t} + C_2 e^{-\sqrt{2}\,t} + 2\sin t + 3\cos t$

$\therefore$ **x &y required solution.**

Example 106: **Solve the simultaneous equations**

$$\frac{dx}{dt} = 2y, \quad \frac{dy}{dt} = 2z, \quad \frac{dz}{dt} = 2x$$

Solution: Given D. E. is $\frac{dx}{dt} = 2y$ ... ... (i)

$\frac{dy}{dt} = 2z$ ... ... (ii)

$\frac{dz}{dt} = 2x$ ... ... (iii)

Differentiating equn(1)w. r. t. 't'

$\frac{d^2x}{dt^2} = 2\frac{dy}{dt} = 2(2z) = 4z$ $\{\because$ from equation (ii)

Again differentiate above equation wrt. t'

$$\frac{d^3x}{dt^3} = 4\frac{dz}{dt} = 4(2x) = 8x \qquad\qquad \{\because \text{from equation (iii)}$$

$$\therefore \frac{d^3x}{dt^3} - 8x = 0 \qquad\qquad \dots \text{This is L. D. E. with constant coefficient}$$

In symbolic form: $(D^3 - 8)x = 0$

$\therefore$ C. S. = C. F. + P. I.

i) To find C. F.

$\therefore (D - 2)(D^2 + 2D + 4) = 0$

$$D = 2, \ D = \frac{-2 \pm \sqrt{4 - 16}}{2} \ ; \qquad \therefore D = 2, \ -1 \pm i\sqrt{3}$$

$\therefore$ C. F. $= C_1 e^{2t} + e^{-t}(C_2\cos\sqrt{3}\,t + C_3\sin\sqrt{3}\,t)$

ii) P. I. = 0

iii) $\therefore$ **C. S. is** $\quad x = C_1 e^{2t} + e^{-t}(C_2\cos\sqrt{3}\,t + C_3\sin\sqrt{3}\,t)$ $\qquad$ ... ... (a)

From equn(i)we have, $\quad y = \dfrac{1}{2}\dfrac{dx}{dt}$

$$\therefore y = \frac{1}{2}\frac{d}{dt}\left[C_1 e^{2t} + e^{-t}(C_2\cos\sqrt{3}\,t + C_3\sin\sqrt{3}\,t)\right]$$

$$= \frac{1}{2}\left[2C_1 e^{2t} - e^{-t}(C_2\cos\sqrt{3}\,t + C_3\sin\sqrt{3}\,t) + e^{-t}(-C_2\sqrt{3}\sin\sqrt{3}\,t + C_3\sqrt{3}\cos\sqrt{3}\,t)\right]$$

i. e. $y = C_1 e^{2t} + \dfrac{1}{2}e^{-t}\left[(\sqrt{3}C_3 - C_2)\cos\sqrt{3}t - (C_3 + \sqrt{3}C_2)\sin\sqrt{3}t\right]$ ... ... (b)

From equn(ii)we have, $\quad z = \dfrac{1}{2}\dfrac{dy}{dt}$

$$\therefore z = \frac{1}{2}\frac{d}{dt}\left\{C_1 e^{2t} + \frac{1}{2}e^{-t}\left[(\sqrt{3}C_3 - C_2)\cos\sqrt{3}t - (C_3 + \sqrt{3}C_2)\sin\sqrt{3}t\right]\right\}$$

$$= \frac{1}{2}\left\{2C_1 e^{2t} + \frac{(-1)}{2}e^{-t}\left[(\sqrt{3}C_3 - C_2)\cos\sqrt{3}t - (C_3 + \sqrt{3}C_2)\sin\sqrt{3}t\right]\right.$$

$$\left. + \frac{e^{-t}}{2}\left[-\sqrt{3}(\sqrt{3}C_3 - C_2)\sin\sqrt{3}t - \sqrt{3}(C_3 + \sqrt{3}C_2)\cos\sqrt{3}t\right]\right\}$$

i. e. $z = C_1 e^{2t} - \dfrac{1}{2}e^{-t}\{(\sqrt{3}C_2 - C_3)\sin\sqrt{3}\,t + (C_2 + \sqrt{3}C_3)\cos\sqrt{3}\,t\}$ $\quad$ ... ... (c)

Hence the equn(a), (b)&(c)taken together gives the required solution.

Exercise

1: Solve the following differential equations

1) $y'' - 2y' + 10y = 0, y(0) = 4, y'(0) = 1$ $\qquad$ 3) $(D - 1)^2(D + 1)^2 y = e^x + \sin^2\dfrac{x}{2} + x$

2) $\dfrac{d^3y}{dx^3} - 3\dfrac{d^2y}{dx^2} + 3\dfrac{dy}{dx} - y = 0$

4) $(D^2 + 6D + 9)y = \dfrac{1}{x^3} e^{-3x}$

5) $(D^4 + 2D^2 + 1)y = x^2\cos x$

6) $(D^2 - 4D + 3)y = \sin 3x.\cos 2x$

7) $\dfrac{d^2y}{dx^2} - 2\dfrac{dy}{dx} + y = xe^x\sin x$

8) $x^2\dfrac{d^2y}{dx^2} - x\dfrac{dy}{dx} + y = x.\log x$

9) $(D^3 - 5D^2 + 7D - 3)y = e^{2x}\cosh x$

10) $(D^2 - 1)y = x\sin x + (1 + x^2)e^x$

2: Solve by the method of variation of parameters (VOP)

1) $\dfrac{d^2y}{dx^2} - 3\dfrac{dy}{dx} + 2y = \dfrac{1}{1 + e^{-x}}$ 3) $\dfrac{d^2y}{dx^2} - 2\dfrac{dy}{dx} = e^x\sin x$

2) $y'' - 2y' + 2y = e^x\tan x$ 4) $\dfrac{d^2y}{dx^2} + y = \dfrac{1}{1 + \sin x}$

3: Solve by the method of undetermined coefficients

1) $\dfrac{d^2y}{dx^2} - 5\dfrac{dy}{dx} + 6y = e^{3x} + \sin x$

2) $(D^2 - 2D + 3)y = x^3 + \cos x$

3) $(D^2 - 2D)y = e^x.\sin x$

4: Solve the following differential equations by Chauchy's Method

1) $x\dfrac{d^2y}{dx^2} - \dfrac{2y}{x} = x + \dfrac{1}{x^2}$

2) $x^3\dfrac{d^3y}{dx^2} + 3x^2\dfrac{d^2y}{dx^2} + x\dfrac{dy}{dx} + y = x + \log x$

3) $x^2y'' + xy' + y = 2\cos^2(\log x)$

4) $x^2\dfrac{d^2y}{dx^2} + 2x\dfrac{dy}{dx} - 12y = x^3\log x$

5: Solve the following differential equations by Legendre's Method

1) $(x - 1)^3\dfrac{d^3y}{dx^3} + 2(x - 1)^2\dfrac{d^2Y}{dx^2} - 4(x - 1)\dfrac{dy}{dx} + 4y = 4\log(x - 1)$

2) $(1 + x)^2\dfrac{d^2y}{dx^2} + (1 + x)\dfrac{dy}{dx} + y = \sin[2\log(1 + x)]$

3) $(3x + 2)^2\dfrac{d^2y}{dx^2} + 5(3x + 2)\dfrac{dy}{dx} - 3y = x^2 + x + 1$

6: Solve the following simultaneous equations

1) $\dfrac{dx}{dt} + y = \sin t,\ \dfrac{dy}{dt} + x = \cos t;$ given that $x = 2$ and $y = 0$ when $t = 0$

2) $\dfrac{dx}{dt} - 7x + y = 0,\ \dfrac{dy}{dt} - 2x - 5y = 0$

3) $(D - 1)x + Dy = 2t + 1,\ (2D + 1)x + 2Dy = t$

4) $(D + 1)x + (2D + 1)y = e^t,\ (D - 1)x + (D + 1)y = 1$

5) $t\dfrac{dx}{dt} + y = 0,\ t\dfrac{dy}{dt} + x = 0; x(1) = 1, y(-1) = 0$

Answers

1: 1. $y = e^x(4\cos 3x - \sin 3x)$ 2. $y = (c_1 + c_2x + c_3x^2)e^x$

3. $y = (c_1 + c_2x)e^x + (c_3 + c_4x)e^{-x} + \dfrac{1}{2} - \dfrac{1}{8}\cos x + \dfrac{x^2}{8}e^x + x$

4. $y = (c_1 + c_2x)e^{-3x} + \dfrac{1}{2x}$

5. $y = (c_1 + c_2x)\cos x + (c_3 + c_4x)\sin x - \dfrac{1}{4}\left[\left(\dfrac{x^2}{12} - \dfrac{3x^2}{4}\right)\cos x - \dfrac{x^3}{3}\sin x\right]$

6. $y = c_1e^x + c_2e^{3x} + \dfrac{1}{884}(10\cos5x - 11\sin2x) + \dfrac{1}{20}(\sin x + 2\cos x)$

7. $y = (c_1 + c_2x)e^x - e^x(x\sin x + 2\cos x)$

8. $y = x(c_1 + c_2\log x) + \dfrac{x}{6(\log x)^3}$ 9. $y = (c_1 + c_2x)e^x + c_3e^{3x} + \dfrac{1}{8(xe^{3x} - x^2e^x)}$

10. $y = c_1e^x + c_2e^{-x} - \dfrac{1}{2(x\sin x + \cos x)} + \left(\dfrac{xe^x}{12}\right)(2x^2 - 3x + 9)$

2: 1. $y = (e^x + e^{2x})\log(1 + e^x) + (c_1 - 1 - x)e^x + (c_2 - x)e^{2x}$

2. $y = e^x(c_1\cos x + c_2\sin x) - e^x\cos x\log(\sec x + \tan x)$

3. $y = c_1 + c_2e^{2x} - \dfrac{1}{2}e^x\sin x$

4. $y = c_1\cos x + c_2\sin x + \sin x\log(1 + \sin x) - x\cos x - 1$

3: 1. $y = c_1e^{2x} + c_2e^{3x} + xe^{3x} + \dfrac{1}{10(\sin x + \cos x)}$

2. $y = e^x(c_1\cos\sqrt{2}x + c_2\sin\sqrt{2}x) + \dfrac{1}{27(9x^3 + 18x^26x - 8)} + \dfrac{1}{4(\cos x - \sin x)}$

3. $y = c_1 + c_2e^{2x} - \dfrac{1}{2}e^x\sin x$

4: 1. $y = c_1x^2 + c_2x^{-1} + \dfrac{1}{3\left(x^2 - \frac{1}{x}\right)\log x}$ .5. $u = \dfrac{kr}{8(a^2 - r^2)}$

2. $y = c_1x^{-1} + \sqrt{x\left[c_2\cos\left[\left(\sqrt{\dfrac{3}{2}}\right)\log x\right] + c_3\sin\left[\left(\sqrt{\dfrac{3}{2}}\right)\log x\right)\right] + \dfrac{1}{2}x + \log x}$

3. $y = c_1x^{-2} + x[c_2\cos(\sqrt{3}\log x) + c_3\sin(\sqrt{3}\log x)] + 8\cos(\log x) - \sin(\log x)$

4. $y = c_1x^3 + c_2x^{-4} + \dfrac{x^3}{98}\log x(7\log x - 2)$

5: 1. $y = c_1(x - 1) + c_2(x - 1)^2 + c_3(x - 1)^{-2} + \log(x + 1) + 1$

2. $y = c_1\cos\log(1 + x) + c_2\sin\log(1 + x) - \dfrac{1}{3}\sin[2\log(1 + x)]$

3. $y = c_1(3x + 2)^{\frac{1}{3}} + c_2(3x + 2)^{-1} + \dfrac{1}{27}\left[\dfrac{1}{15(3x + 2)^2} + \dfrac{1}{4(3x + 2)} - 7\right]$

6: 1. $x = e^t + e^{-t}$, $y = e^{-t} - e^t + \sin t$

2. $x = e^{6t}(c_1\cos t + c_2\sin t), y = e^{6t}[(c_1 - c_2)\cos t + (c_1 + c_2)\sin t]$

3. $x = -t - \dfrac{2}{3}, y = \dfrac{1}{2}t^2 + \dfrac{4}{3}t + c$ 5. $x = \dfrac{1}{2}\, t - \dfrac{1}{t}, y = \dfrac{1}{2}\left(-t + \dfrac{1}{t}\right)$

4. $y = c_1 e^x + c_2 e^{-2x} + 2e^{-x}, \quad z = 3c_1 e^x + 2c_2 e^{-2x} + 3e^{-x}$

www.ingramcontent.com/pod-product-compliance
Lightning Source LLC
Chambersburg PA
CBHW062343300326
41947CB00012B/1196